Science Networks · Historical Studies
Volume 20

Edited by Erwin Hiebert and Hans Wussing

Springer Basel AG

George Boole

Selected Manuscripts on Logic and its Philosophy

Editors:
Ivor Grattan-Guinness
Gérard Bornet

Springer Basel AG

Editors:

Ivor Grattan-Guinness
Professor of the History of
Mathematics and Logic
Middlesex University
Queensway
Enfield
Middlesex EN3 4SF
Great Britain

Dr. phil.-hist. Gérard Bornet
Philosophisches Seminar
Université Miséricorde
1700 Fribourg
Switzerland

Published with support of the Swiss National Science Foundation.

Library of Congress Cataloging-in-Publication Data

Boole, George. 1815–1864
   George Boole : selected manuscripts on logic and its philosophy /
   [edited by] Ivor Grattan-Guinness, Gérard Bornet.
   -- (Science networks historical studies ; v. 20)
   Includes bibliographical references and index.
   ISBN 978-3-0348-9805-8   ISBN 978-3-0348-8859-2 (eBook)
   DOI 10.1007/978-3-0348-8859-2
   1. Logic, Symbolic and mathematical. 2. Mathematics--philosophy. 3. Boole,
   George, 1815–1864.   I. Grattan-Guinness, I.   II. Bornet, Gérard.
   III. Title. IV. Series.
   QA9.2.B68   1997
   511.3--dc21

Deutsche Bibliothek Cataloging-in-Publication Data
George Boole – selected manuscripts on logic and its
philosophy / Ivor Grattan-Guinness/Gérard Bornet. - Basel ;
Boston ; Berlin : Birkhäuser, 1997
   (Science networks ; Vol. 20)

   ISBN 978-3-0348-9805-8
NE: Grattan-Guinness, Ivor; Bornet, Gérard; GT

© 1997 Springer Basel AG
Originally published by Birkhäuser Verlag in 1997
Softcover reprint of the hardcover 1st edition 1997

Cover Design: Micha Lotrovsky, Therwil, Switzerland
Printed on acid-free paper produced from chlorine-free pulp. TCF ∞

ISBN 978-3-0348-9805-8

9 8 7 6 5 4 3 2 1

Contents

Editors' Introduction

Part A
The Nature of Logic and the
Philosophy of Mathematics

Part B
The Philosophical Interpretation
of a Theory of Logic

Chapter VI

Chapter VII

Part C
"The Philosophy of Logic" – A Sequel to "The Laws of Thought"

Chapter VIII

Chapter IX

Chapter X

Part D
Miscellaneous Matters,
Letters and Fragments

Editors' Introduction

Part 1: Boole's quest for the foundations of his logic

Ivor Grattan-Guinness

§1 Our purpose

The adjective "boolean", sometimes even "Boolean", is known to mathematicians for a powerful algebra, to electrical engineers and computing specialists who apply it widely in parts of their work, to philosophers as a component of logic, and to historically-minded people as one George Boole (1815-1864), who formed an algebra of logic from which we enjoy these modern fruits. This book publishes some of the manuscripts in which he tried but failed to find a satisfying philosophy to underpin that logic, and conform with the elements of philosophy and psychology which he felt himself to have addressed; almost all appear for the first time.

After a short recollection of Boole's life and work (§2), this introduction continues with the story of the manuscripts, in particular of the collection from which our edition is drawn (§3); the organisation of our selection is described in §4. Then, following brief surveys of the revival of interest in logic in Britain from the 1820s (§5) and of Boole's mathematical background (§6), we describe the main phases of his logical career. He wrote his first book, *A mathematical analysis of logic* (hereafter "*MAL*") of 1847 (§6-7). Then a period of rethinking over seven years (§8) culminated in his second and more substantial book, *An investigation of the laws of thought* ("*LT*") of 1854, which included applications of his logic to probability theory (§9). After that he wrote many manuscripts, especially to write a third book on "The Philosophy of Logic" ("PL") (§10-12). We end with some remarks on four figures who did *not* influence Boole as one might expect (§13), and a short review of the impact of his publications up to the end of the century (§14). Throughout we give more space to his mathematical details and procedures, because he tended to assume them as known in the manuscripts, where he covered the philosophical and psychological aspects much more amply.

§2 Life and career

George Boole was born on 2 November 1815 in Lincoln, the eldest of four children of a local tradesman John Boole (1777-1848) and his wife Mary Ann (1780-1854).[1] He adopted a love of learning from his father to such an extent that already in his teens he taught at schools in Lincoln and nearby; in his 20th year he even opened his own school. He also learnt classical and modern languages and wrote poetry, and taught himself mathematics (and also French and German) to a level which permitted him to engage in research in mathematics. He began publishing in 1841, first and frequently in *The Cambridge Mathematical Journal*, which had been founded two years earlier by D. F. Gregory (1813-1844), his initial mathematical mentor and inspiration (§7).

Although this early work gained considerable attention, Boole had to continue in school-teaching until the Queen's University was proposed in 1845, as a secular institution for higher education in Ireland. After delays caused by the potato famine, and Catholic opposition, Queen Victoria formally opened the constituent colleges, including Queen's College, Cork, in 1849, and Boole took up then his post as the founder Professor of Mathematics. While remembered as a conscientious and at times inspiring teacher, he was not able to find any distinguished followers among his student clientele. Never really happy with life in Ireland, he was recognised only locally; in particular, he had very little contact with W. R. Hamilton (1805-1865) up north in Dublin, the only other mathematician in the island of comparable calibre, even though they were both pioneering new algebras. Further, the Royal Irish Academy did not elect him to their throng. However, he received honorary doctorates from Trinity College Dublin in 1851 and Oxford University eight years later, and was made Fellow of the Royal Society in 1857. For likenesses there survive a very few sketches, and one photograph taken in London probably in 1860 and often reproduced; when printed the correct way round he is looking to the viewer's left (Figure 1).

Throughout his career Boole published steadily, in two areas which were tightly linked, as we shall see in §6 and §7. He started with the differential and integral calculus, where he became a major figure in certain algebraic theories popular in England and Ireland. He also wrote two well-received textbooks: *A Treatise on Differential Equations* (*1859a* in the bibliography),

Figure 1: The well-known photograph of Boole, taken in
London 1864 (Reproduction D. MacHale)

and *A Treatise on the Calculus of Finite Differences* in the following year (*1860a*).

Boole's other area was logic, which he publicised principally in *MAL* (*1847a*) and *LT* (*1854a*), and in some papers. Today the latter book is recognised as more authoritative, but at the time it gained less attention than *MAL*. To accentuate the historical irony, in his lifetime and for years afterwards his work in logic was much *less* well recognised than were his contributions to the calculus, where his methods and especially textbooks were used. Doubtless aware of the apathy over the logic, he attempted at various times, especially soon after the publication of *LT*, to write the non-technical PL.

In 1855 Boole married Mary Everest (1832-1916), daughter of the Reverend Thomas Everest and niece of Sir George Everest, after whom the famous mountain is named. They produced five daughters, four of whom had varied and distinguished careers of their own: two mathematicians, a chemist and a novelist. One of them will appear in §3 as an archivist. Mary assisted her husband with his two textbooks on differential equations, as the first reader for intelligibility and also as checker of some of the exercises. After his death she proselytised Boole's philosophy, especially its application to educational questions, both in teaching and in print. Although she was regarded as a crank, her understanding of his ideas on logic and education were basically sound (Laita *1980a*): scattered around her rather verbose collected works are some valuable hints and data. In particular, the title of the short piece "Mount Carmel in London", and its initial publication in the journal *The Crank* in 1904 (Boole, M. *1931a*, 75-79), should not deter readers from its merit.

Like his wife, Boole was deeply religious. According to her, when a youth he came to the idea that a universe was central to knowledge and revelation as in "the Great All stressed as a unity in the Scriptures", and monism in the Jewish faith (Boole, M. *1931a*, 473, 951-952), and thereafter he affirmed ecumenism. He implicitly exhibited his position in *LT* by devoting ch. 13 to sophisticated logical analyses of propositions due to Samuel Clarke and Benedict Spinoza concerning the necessary existence of "*Some one unchangeable and independent Being*" (p. 192). However, he was a Dissenter, and at a time when such movements were gaining considerable ground in Britain against strong resistance from the established Trinitarian Church of England. He alluded to his stand in print very discreetly a few lines from the end of *LT*, where he mentioned "the Father of Lights" and finished off with some enigmatic lines about the bearing of religious belief upon his logic.

Ecumenism was very important to the philosophy of Boole's logic; for its Universe 1 symbolised the Godhead standing over and lighting the factions into which the Christian Church was split. He was especially admiring of Frederick Denison Maurice (1805-1872), who advocated ecumenism with great force in mid century and so was dismissed from his chair at King's

Figure 2: Boole's house at Ballintemple, where he died
(Photograph I. Grattan-Guinness)

College London. Boole spent parts of several summer vacations in England in his last years, studying in London libraries and also attending Maurice's Sunday services.

In similar reticence, Boole kept his political views to himself: he offered no utterances on the condition of the Ireland in which he lived after the famine. But in a manuscript discussion of words, written in the revolutionary year 1848, he embodied his liberal aspirations (page 5 below).

Boole died as a result of foolishly walking the three miles from home to university without proper protection from a rain storm, so probable an event in Cork. The ensuing pleuro-pneumonia took him away in his 50th year on 8 December 1864; he was buried at Blackrock, near Cork and the family residence in Ballintemple (Figures 2 and 3).

The popular *Illustrated London News* published a short obituary by the Vice-President of the College (Ryall *1865a*) together with a fine sketch (Figure 4). Boole's former friend the teacher and mathematician Robert Harley (1828-1910) published a fine long eulogy *1866a*, and condensed from it a routine obituary *1867a* for the Royal Society. The Scottish educationalist and Shakespeare scholar Samuel Neil (1825-1901) also wrote an excellent piece *1865a* in his own journal, *The British Controversialist*.[2] In Canada there

Figure 3:
Grave of Boole at Blackrock,
with his biographer D. MacHale
(Photograph I. Grattan-Guinness)

appeared a lengthy appreciation of *LT* (Young *1865a*), together with a short notice (Anonymous *1865a*). But no obituaries were written by leading mathematicians.

In addition, and exceptionally for a mathematician, two stain-glass windows were erected in Boole's memory. One was inserted around 1867 in the Great Hall of Queen's College Cork; he is shown in one of the ten panels (Figure 5) portraying various world luminaries. The other was inserted into Lincoln Cathedral two years later, thanks to a subscription raised by fellow townspeople.

§3 The fate of Boole's Nachlass

After Boole's death and burial, his widow moved the family to London, where Maurice found her a post at his Queen's College for women's education in Harley Street. Although the government awarded her a Civil List pension of £100 per year, lack of money presumably forced her to sell his library. No

Figure 4:
Engraving of Boole,
published with Ryall's
obituary *1865a*

list of it survives, but it was probably not large; it seems likely that most of his wide reading was done in libraries, or at least thanks to their holdings.

As well as teaching and applying George's ideas to education, Mary began sorting his manuscripts with the help of her third daughter Alicia (1860-1940). Mathematical colleagues saw them: William Spottiswoode (1825-1883) fulfilled her request to send some of the mathematical ones to the Royal Society, and Isaac Todhunter (1820-1884) quickly formed those which Boole had prepared for a second edition of the book on differential equations into a "supplementary volume" of the (Boole *1865a*), the edition itself was basically just a corrected reprint.

In addition, Augustus De Morgan (1806-1871), Boole's most eminent contemporary in the algebra of logic (§13) and a regular correspondent, looked through the logical manuscripts. He felt in his report *1867a* that only one was fit for printing: Boole *1868a* on numerically definite syllogisms (a topic initiated by De Morgan himself in his *1847a*, ch. 8), seemingly written around 1850.

Mrs Boole and Alicia sorted out the mathematical from the logical manuscripts, organised each part into lettered series, and transcribed by hand many non-symbolic logical texts and also letters; they also gathered together a group of notebooks. Eventually she arranged to give them (hereafter "the collection") to the Royal Society in 1873. However, family interest in them had not ceased. In particular, in the 1880s Alicia, then living in or near Liverpool, teamed up with fellow townsman H. J. Falk, mathematician by inclination and barrister by profession. They formed a small group to study Boole's work; it included an actuary called Walter Stott, whom Alicia married in 1890. In order to prepare a new edition of *LT*, which had became very rare, they asked to borrow back the collection from the Royal Society in 1889. The Council minutes of 20 June duly record that "Leave was granted to Miss Boole to borrow the MSS. of the late Prof. Boole, and to make extracts from the same" (Royal Society *1893a,* 262).

Alicia and her helpers looked in detail at the logical part; they tried to determine which manuscripts were written before and which after *LT*, and to pick out those of the latter destined for the succeeding PL. They organised the collection further into various lettered and numbered sections, and prepared typescripts of several logical manuscripts and some letters. They returned it to the Royal Society in 1896; Council minutes of 29 October (Royal Society *1899a,* 293-294) record that

> The Boole MSS. presented to the Society by the widow of the late Prof. George Boole and lent to Miss Boole in 1889, for the purpose of making extract therefrom, have now been returned, the whole of the MSS having been collated and arranged in order, and type-written copies of several of them furnished for the Society's Library by Mr. H. J. Falk, on behalf of the Boole family. It is proposed to proceed now with the binding of the collection.

The Society also resolved to thank Falk "for the trouble he has taken in arranging and copying the Boole manuscripts". Mercifully, the binding was never effected; in addition, no publications came from all this effort, although Mrs Boole still published on (her version of) his theories from time to time.[3]

Then, in 1905, enter Bertrand Russell (1872-1970), surprisingly and probably also surprised. In his early thirties, he was deeply engaged in a vast logical enterprise with A. N. Whitehead (1862-1947) of a quite different kind from Boole's (§14) which was to yield their three-volume *Principia mathematica* (1910-1913). Perhaps not appreciating the great differences between this and her husband's enterprises, Mrs. Boole seems to have thought of Russell as a suitable editor. Once again nothing eventuated; but parts of the related correspondence survives in Russell's *Nachlass.* Falk's son Oswald Toynbee[4] told Russell on 18 February 1905, not quite accurately, that

> Mrs. Boole after George Boole's death gave a large collection of his unpublished papers to the Royal Society absolutely. The more strictly mathematical portion of the papers was looked into by Todhunter and I believe De Morgan, the result of their work being

Figure 5:
Panel showing Boole in the
Memorial Window in the Great
Hall of University College, Cork
(Photograph I. Grattan-Guinness)

the supplementary volume of the Differential Equations. My father thinks that they extracted nearly all that was valuable of this nature.

There were however in addition many papers which appeared to be connected with "The Laws of Thought", and with the permission of the Royal Society my father had all the papers out for some years, and in conjunction with Alice Boole (Mrs. Stott) he produced some sort of order in the logical portion of the papers. They are I believe the beginnings of a book designed to render the ideas of the "Laws of Thought" more intelligible to the non-mathematical reader. The papers after arrangement were copied, (3 typewritten copies were made of which my father has one) and the originals with a copy given back to the Royal Society where they now lie. The Royal Soc[iet]y asked my father if he had any recommendation to make re publishing, but apparently he did not like to judge for himself, and no one else has taken the matter in hand. The papers may be worth publishing and apparently all that is required is the judgement of somebody qualified to decide in such a matter. If you would like to see the papers (or have not already seen them) my father would be glad to send you his copy after obtaining Mrs. Boole's permission.

The old lady is, I believe, rather a difficult person to deal with but I expect there would be no opposition.

You will see that I rather overestimated the work which remains to be done. I had heard my father talk of the confusion of the papers but did not know that anything had been done to put them in order or that they had been restored to the Royal Society.

Russell corresponded with his friend the logician Louis Couturat (1868-1914) about the request at this time, correctly considering him to be a far more appropriate candidate as editor; among other work, Couturat had already written on the similarities between Boole's and Leibniz's logics (§13) (*1901a*, 344-354, 385-387). On 7 February he wrote to Russell, regarding it as "a great honour for me to be charged with the edition of these manuscripts, on condition that they are worth the effort (that is to say that they are not simple drafts or copies of what Boole had published) and on condition that they do not pass my competence such as treating probability theory."[5] In reply on 9 February, Russell offered to translate into English any editorial material that Couturat might write.

However, after hearing nothing from Falk, he wrote on 13 October from his home at Bagley Wood near Oxford:

I have never heard anything further from you about the Boole MSS in your father's possession. As it occurred to me that perhaps you were kept from writing by not knowing my address, I am writing to say that the above is my permanent address. I heard from M. Couturat that if the MSS proved interesting he would probably undertake to edit them; & I feel no doubt that he is the most suitable person living for the purpose.

This initiative inspired further recollections by H. J. Falk to his son on 18 October, which were forwarded to Russell:

My dear Oswald,

It would be well to make clear to Mr. Russell that the Boole MSS are not in my possession or control. They are at the Royal Society to whom they were committed by Mrs. Boole on her husband's death 40 years ago.

Any control I have is solely by courtesy through my having (with Mrs. Boole's consent and permission of the Royal Society Council) collated and arranged the chaotic mass of papers. I then had the MSS typed, presented one copy to the Royal Society and retained one. I have for 20 years wished to see the "Laws of Thought", reprinted with some of these MSS and perhaps a biography added, and would gladly give Mr. Russell and his friend any assistance of which I am capable to that end. It would be a real public service and I regret that there are objections in the family, which so far I have never overcome. I feel bound to respect them, not merely on friendly grounds. Mrs. Boole holds various metaphysical views as to Boole's Logic, based chiefly on Gratry,[6] which I have never been quite able to follow. But I cannot venture for that reason to conclude that they are without foundation.

She must remember much of her husband's thought, which no one now can know. These views are not shared by her very able daughters and especially not by Mrs. Stott, who has most of her father's mathematical and logical genius. Now Mrs. Stott worked very hard with me on the MSS, and I cannot disregard her wishes. To put it briefly, her fear (in which the others concur) is that, if the L.T. or M.S.S. be printed by anyone, Mrs. Boole will not only disown any connection with the new edition but will attack it as an inadequate presentation of Boole's thought. Hence they would prefer nothing done during the old lady's life. A few words as to the M.S.S. — In the ten years between the L.T. (1854) and his death (1865) Boole was continually working at a sequel to it. The important part of the MSS consists of those efforts. His great mind sought to lift the Principles and Methods of Thought even further than De Morgan and Jevons[7] seemed to see, — into a clear region beyond the schoolmen and formalists, whether logical or mathematical. Thus the beauty and power of those MSS are, if I am any judge, too precious to be buried for ever.

Russell responded to this envoi on 25 October as follows:

Many thanks for your letter, and for the letter from your father enclosed. It would give me the greatest pleasure to meet your father if it can be arranged. I hope if anything brings him to Oxford, he will come here at any time that suits him; otherwise, we might meet in London.

From his letter I see that the objections to printing during Mrs. Boole's lifetime are very strong. I have always taken it for granted that her views of Boole's opinions must be wrong, particularly since I read Gratry about a year ago: Gratry seemed to me not nearly good enough. I should therefore be quite in sympathy with the feeling that it would be better to wait.

I am afraid from your father's letter that he thinks I am willing to edit the MSS myself, in case at any time they are to be edited. This is hardly the case. I have at all times a great deal of original work that I want to do, & I should hardly be willing to give up so much time. But M. Couturat is more learned than I am: he has done an *admirable* edition of Leibniz's unpublished MSS on Symbolic Logic [Couturat *1903a*]; he is willing to undertake the task and is far better qualified for it than I should be. I should of course be glad to assist him in any way, particularly as regards language. But *he* would be the editor.

Despite Couturat's willingness to act, nothing happened in the end. However, a few years later the mathematician and historian of mathematics Philip Jourdain (1879-1919), who had taken Russell's course in mathematical logic while an undergraduate at Cambridge in 1901-1902, became interested in Boole, as part of a general study of the recent development of symbolic logic. He wrote an article *1910a* on Boole, borrowing from the Stotts copies of various materials, and also from Boole's former friend the mathematician and teacher Robert Harley (1828-1910) a bound volume containing Boole's letters to him and copies of various others; sadly, this volume has never been seen since. Jourdain's only other act was to publish with the Open Court Company in 1916 a new edition of *LT*; however, he contributed merely a two-page introduction, corrections of errata, and a (very welcome) index, and did not include any manuscripts. Published as "Volume 2", it was intended to be preceded by an edition of Boole's papers on logic; but he seemed not to have started on this project before his early death. His manuscripts were passed by his widow to A. E. Heath and J. M. Child as literary executors (Rhees *1955a*); but most have disappeared.

Thus ended serious Boole scholarship until the late 1940s. The centenary of *MAL* in 1948 stimulated a survey Kneale *1948a* of Boole's life and work in *Mind*; and a photoreprint of the book that year aroused some interest, such as an article Prior *1949a* on propositions in Boole. Then four years later two British philosophers used the collection. Mary Hesse quoted a few passages in an article *1952a*; and at the suggestion of Jourdain's literary executor A. E. Heath, Rush Rhees produced Jourdain's missing volume with a selected edition Boole *1952a* of published and unpublished writings on logic and probability, prefaced by his own informative "Note in editing". (Jourdain's edition of *LT* was reprinted with it, for the second time.) Both these works contain short quotations from some of the manuscripts included here. In 1954 the Royal Irish Academy celebrated the centenary of the publication of *LT* with a special meeting and proceedings; Rhees *1955a* and Feys *1955a* will be cited from the latter.

After another pause, this time of 20 years, several historians used the collection. Susan Wood worked on it, in the course of preparation of a doctoral dissertation *1976a* on Boole's theory of propositional forms. The most noteworthy later results include Desmond MacHale's biography *1985a*, and Maria Panteki's exploration of the connection between algebras and logic in Boole in her doctoral thesis *1992a* on this theme in Britain over the period 1810-1870.

The collection is currently stored in nine boxes, together with a separate set of notebooks. Some general remarks on our editing and dating are given in the short set of "Remarks" following this introduction, with the necessary minimum of grim details supplied in the "Textual notes" which follow the transcriptions.

As with many figures, Boole's full *Nachlass* has become somewhat scattered. In particular, quite a few documents found their way to the USA, seemingly in the family of Boole's elder daughter Mary Ellen (1856-19??), and were sold at auction to Cork University Library in 1983. They include an unpublished biography by his sister Mary, quite a few letters and notebooks, and some material from his school-teacher days. While we have consulted them, none has been included in our selection.

§4 Organisation and explanation of the selected manuscripts

Boole seems to have tried again and again to articulate his philosophy of logic, usually starting each time from scratch (as he told De Morgan in a letter quoted in §11). Only rarely did he produce a substantial text; quite often the outcome was more notes or another work-plan. He never dated a manuscript, although occasionally he indicated a time by reference to *MAL* or *LT*; and his handwriting did not noticeably change over the years. Another complication is that sometimes he wrote essays or plans on logic on the versos of discarded folios of mathematical remarks or calculations, and some of these sequences have been ordered for mathematical rather than logical continuity.

Thus our selection has been made from a scattered and chronologically unclear textual universe. It comprises around 40% of Boole's unpublished manuscripts on logic. We hope that it provides a wide and characteristic cross-section of his meditations; it complements a few manuscripts which were included by Rhees in his edition Boole *1952a*. We have divided it into four Parts lettered A-D, and also into Chapters numbered I-XVII. Most Chapters comprise one integral manuscript, and some contain Boole's own chapters or at least references to them; so to avoid confusion we refer to his (and anyone else's) "chapter(s)". In the same spirit, this book has "page(s)" while all others have "p(p)". Unless indicated by our square brackets, the title of each Chapter is Boole's.

The orders of, and in, Parts A-C are hopefully chronological, at least coarsely so. There seems to be nothing before *MAL*; Part A is placed between *MAL* and *LT*, and is surveyed in §8. Part B contains manuscripts written after *LT* (§10), and Part C those which seem to have been specifically intended for the third book (PL) on the philosophy of logic (§11); the division between Parts follows the intention of Alicia Boole and Falk a century ago, but we cannot claim it to be perfect. Those sections of all these writings relating most closely to mathematics are highlighted in §12, and from them a conjecture is made about his failure to produce PL. Part D contains two manuscripts of uncertain date (Chapters XII and XV) whose contents place them rather outside Parts B and C, some interesting notes and undated fragments, and a few letters; they are all noted in §8 and §10-§12 as appropriate.

After the "Textual Notes" comes the bibliography, which combines the items cited in this introduction, a few other historical writings of note which have not been cited, and the works mentioned or alluded to in the manuscripts and identified in the Notes. An index of names and subjects completes the edition.

§5 The renaissance of logic in Britain

Boole's contributions to logic came at a time when the discipline was experiencing a remarkable revival after a long period of quietude. While he seems not to have been profoundly influenced by recent developments when forming his own approach, he drew upon them, especially in his manuscripts. Some main details, therefore, should be noted.

The study of logic in Britain was then in a peculiar state. The classical tradition, based upon inference in syllogistic logic, was still in place. But for a long time an alternative tradition had been developing; inspired by John Locke (1632-1704) and continued in some ways by the Scottish Common-Sense philosophers of the late 18th century. Critical of syllogistic logic, especially for its narrow concern with inference, adherents of the new tradition sought a broader foundation for logic in the facultative capacity of reasoning in man, and included topics such as truth and induction which we might now assign to the philosophy of science. Showing more sympathy to the role of language in logic than had normally obtained among the syllogists, they laid emphasis on signs as keys to logical knowledge; indeed, Locke used the word "semiotics" in the fourth Book of his *Essay concerning human understanding* (1690) (Buickerood *1985a*). At the cost of some simplification, we shall refer to this approach as "the sign tradition".

The principal source of this revival of interest can be precisely identified: the *Elements of logic* (1826) by Richard Whately (1787-1863), based upon his long essay on the topic in the *Encyclopaedia metropolitana*, which had been published in two parts in 1823. In the book version he surveyed syllogistic logic in a "Synthetical Compendium" of nearly 80 pages, and used it in a following chapter of equal length on "Fallacies". But he was much taken with the sign tradition, endorsing some of its criticisms of syllogistic logic in his "Introduction" and opening chapter.

Whately's book prompted an astonishing response; he soon produced revised editions (the fourth in 1832), and further ones later. Several figures wrote lengthy reviews or commentaries, or used it in writing textbooks for university use. All figures in logic of that time drew something from him, whether positively or negatively, at least in English-speaking countries.[8] Boole was one reader; so we use Whately as our main source for several issues, although other authors doubtless also influenced him.[9]

The main single innovation was made by George Bentham (1800-1884), in his *Outline of a new system of logic with a critical examination of Dr Whately's "Elements of Logic"* (1827); but it sold very poorly, and so was not widely recognised until the 1850s. He greatly increased the number of forms in syllogistic logic by adding to the standard ones such as "All As are Bs" the types "All/Some/No As are All/Some/No Bs". The name "quantification of the predicate" became attached to this extension during the 1840s (it is the origin of the word "quantification" in logic). A priority row over its genesis was initiated in the mid 1840s by the Scottish philosopher Sir William Hamilton (1788-1856) (his approach to logic is summarised in §11). However, although he had seen Bentham's book, his target was De Morgan (Laita *1980a*). It was this contretemps which inspired Boole to write up his ideas on logic, and to publish them as *MAL* in 1847.

§6   Boole, English algebraist par excellence

Boole's contributions to mathematics and logic were both characteristic of and important for the development of English mathematics in his time; for they were entirely guided by algebras of new kinds, which we consider here. His earliest work extended the notion of the invariant, which was to become a major English industry in the hands of Arthur Cayley (1821-1895) and J. J. Sylvester (1814-1897) (Parshall *1989a*). But his main concern was drawn from the algebraic version of the calculus based upon Taylor series, proposed by J. L. Lagrange (1736-1813), extended in the early 19th century by the Alsatian L. F. A. Arbogast (1759-1803), and continued by a few minor French mathematicians (Grattan-Guinness *1990a*, chs. 3-4).

In this version the operation of differentiating a mathematical function $f(x)$ was represented by the letter "D", so that the derivative $f'(x)$ (to use Lagrange's notation) was written "$Df(x)$". An algebra was developed in which second differentiation was written as the power "D^2", the inverse operation of integration was "D^{-1}", and so on. The resulting theory was used principally to solve differential and difference equations, sum series and evaluate integrals. The results were often correct although the foundations of this algebra were mysterious; for example, as critics pointed out, the orders of differentiation were identified with the corresponding powers of the first $(d^n y/dx^n) = (dy/dx)^n$ for $n \geq 2$).

Inspired partly by Gregory, Boole became a leading practitioner of this theory, not only making new applications but also seeking to explicate the foundations and make the theory more rigorous (Panteki *1992a*, ch. 4). His principal paper, "On a general method in analysis", was submitted to the Royal Society; after wishing to decline it, they published it in their *Philosophical Transactions* as Boole *1844a*, and then gave him their Gold Medal for it. In this paper and related ones, he explored the D-methods to new lengths, applying them to both ordinary and partial differential equations, and examining non-commutative operator functions F and G for which F(D)G(D)≠G(D)F(D). He seems to have hoped that one general method could be found; but his own work revealed the hope as overly optimistic.

Three years later *MAL* appeared, in which Boole gave the first version of his logic. As its title shows, he saw his "mathematical analysis of logic" as *applied mathematics,* a view which he always maintained. In this application he worked with mental acts (called "elective symbols", in the manner common to algebraists of that time of identifying the symbol with the object or operation symbolised); the act of selecting, say, men out of the Universe of humans. He also spoke of the class of men so selected, and interpreted his theory for both operations (mental acts) and objects (classes). "Let us employ the letters X, Y, Z, to represent the individual members of classes [and] The symbol x operating upon any subject comprehending individuals or classes, shall be supposed to select from that subject all the Xs that it contains" (*MAL*,15). The "product xy will represent, in succession, the selection of the class Y, and the selection from the class Y of such individuals of the class X as are contained in it, the result being the class whose members are both Xs and Ys" (p.16). "The equation $y = z$ implies that the classes Y and Z are equivalent, member for member" (p.19).

Boole then set down the basic laws with these mental acts obeyed, in a form closely similar to those for the differential operators in *1844a* (*MAL*, 17-18). Writing π and r for functions of D operating on the mathematical functions q and r, and u, v, x and y for elective symbols, he named and wrote the laws for the two theories as follows;[10] l, m and n are positive integers, with n≥2:

$$\pi r q = r \pi q \qquad \text{commutative law} \qquad xy = yx \qquad (6.1)$$

$$\pi(q+r) = \pi q + \pi r \qquad \text{distributive law} \qquad x(u+v) = xu + xv \qquad (6.2)$$

$$\pi^l \pi^m q = \pi^{l+m} q \qquad \text{index law} \qquad x^n = x \qquad (6.3)$$

§7 Logic in Boole's *Mathematical analysis of logic*

Boole's principal single innovation was law $(6.3)_2$ for logic: it stated that the mental acts of choosing the property x and choosing x again and again is the same as choosing x once, or that the class x taken together with the class

Figure 6:
Painting of the young Boole in
his early thirties (around1845)
(Reproduction D. MacHale)

x over and over again gives the class x. (The possible parentage of this law in Leibniz is considered in §13.) As consequences of it he formed the equations

$$x(1-x) = 0 \qquad \text{and} \qquad x+(1-x) = 1 \qquad\qquad (7.1)$$

which for him expressed respectively the law of contradiction and the law of excluded middle. The various symbols were interpreted as follows: for brevity we mention just the readings as classes. "1" symbolised a "Universe" which "comprehend[s] every conceivable class of objects whether existing or not" (p. 15), and also the corresponding operation (p. 20); but this was rather too Universal, since within this 1 true propositions cannot be distinguished from tautologically true ones. "–" represented the subtraction of a class from one of which it was a part; however, since he never used a symbol for "not", his expression of that law was not complete (Ellis *1863a*). "0" denoted "Nothing" — none too clearly, since it was a class of some sort. In addition, the theory of classes which he used was the extensional part-whole kind, with (for example) x as part of 1. He also defined the conjunction $(x+y)$ of two classes only when x and y had no parts in common: this met with a poor reception (§14). He left implicit quite a few laws; associativity of addition and of multiplication, for example, and those which 0 and 1 obeyed.

One main purpose of Boole's algebra was to take one or several propositions as premises, express them as algebraic equations, and then to relate a selected elective symbol to the others; the resulting proposition was a logical consequence of the premises. The double interpretation of letters as elective symbols or as classes gives the algebra a feel of model theory, although he cannot be read in its modern terms, or even of its pioneers around 1900 (Scanlan *1991a*).

Boole realised that a given set of premises may not have any solutions, or maybe more than one; and to accommodate the latter possibility he introduced "V" to symbolise an indefinite proportion of members of a class, from none through "some" to all of them; the corresponding elective symbol was "v" (p. 21). This notion has often perplexed readers, but it allows him to handle the lack of a cancellation law in his algebra: namely that

$$\text{if } xy = xz, \text{ then it is not necessarily the case that } y = z. \qquad (7.2)$$

In addition, it enabled him to express many propositions in terms of equations rather than inequalities, which would have complicated the algebra considerably. However, he offered no laws which "v" should satisfy, and could not always distinguish between traditional forms of proposition and those involved in the quantification of the predicate (which he did not analyse explicitly); for example, "$vx = vy$" could cover both "Some Xs are Ys" and "Some Xs are some Ys" (pp. 21-22).

Further, contrary to Boole's apparent belief, the solutions found by his methods were not always complete (Corcoran and Wood *1980a*). For example, for the universal affirmative proposition "All Ys are Xs", symbolised as

$$(1-x)y = 0, \qquad \text{he put forward} \qquad y = vx \qquad (7.3)$$

as "the most general solution" (*MAL*, 25); but he should have noticed that $x = 0$ was missing from it, and that it did not hold if $x = 0$ and v was a class such that $vy \neq 0$. It is curious that a logician should seem to slip over the difference between necessary and sufficient conditions (another apparent example arises in the manuscripts at (10.1)), or that a mathematician with a strong interest in singular solutions of differential equations (§12) did not notice analogous problems in his logic. In fact, in this case Boole realised that $v = xy$ was a better version than $(7.3)_1$: so did the Irish mathematician Charles Graves after reading *MAL*, and he told the Royal Irish Academy in a note *1850a*.

A substantial section of *MAL* (pp. 48-59) treated "Of Hypotheticals", propositions composed of two or more propositions A and B linked by a connective, such as "A and B", say, or "If A, then B". His procedure followed the practise of Whately and others to treat them as categorical propositions concerning their constituent propositions; but the resulting algebra was rather strange (Prior *1949a*, Hailperin *1984a*). "1" was now charged to symbolise the "hypothetical Universe" of all circumstances, so that x elected those for which proposition X was true. Then, for example, "the Proposition, either X is not true, or Y is not true, the members being exclusive" because of the restriction on "+", came out as

$$y(1-x)+x(1-y) = 1 \qquad \text{for} \qquad x-2xy+y = 1 \qquad (7.4)$$

(p. 53); the index law $(6.3)_2$ converted it to a quadratic equation, and then factorising yielded solutions. This part of his theory did not endure (§9).

In order to generalise his methods of drawing consequences from premises, Boole devoted the rest of his book to various general expansion theorems and processes of elimination of letters. The theorems were proved by an extraordinary use of Taylor's series with $(6.3)_2$ used to get rid of the powers (pp. 62-63); they took these forms for a function "$\phi(x)$" of one mathematical act, and "$\phi(xy)$" or a function of two of them:

$$"\phi(x) = \phi(0)+\{\phi(1)-\phi(0)\}x = \phi(1)x+\phi(0)(1-x)", \qquad (7.5)$$

$$"\phi(xy) = \phi(00)(1-x)(1-y)+\phi(01)y(1-x)+\phi(10)x(1-y)+\phi(11)xy". \qquad (7.6)$$

The basic components x and $(1-x)$ of (7.5), or xy, ... and $(1-x)(1-y)$ of (7.6), perform like a basis in a space of expansions. Although vector space theory was not yet explicitly in place, linear combinations were well recognised, and Boole soon stated the generalisation appropriate to his algebra on p. 64:

$$"a_1t_1+a_2t_2 \ldots +a_rt_r = 0" \qquad (7.7)$$

for base terms $\{t_j\}$ with coefficients $\{a_j\}$.

Boole obtained the solutions by the usual methods of linear equations adapted to the laws of this algebra. As a result, some of them were expressed as quotients of the functions, with particular values for the coefficients. For example, $(7.3)_1$ was solved for y from (7.5) as

$$"y = \frac{1}{1-0}x+\frac{0}{0-0}(1-x) = x+\frac{0}{0}(1-x)" \qquad (7.8)$$

with "$\frac{0}{0}$" was explained as an alternative to "v" (p. 74), a quotient notation which was to feature prominently in *LT*. Despite the extended method of solution, the same "complete solution" $(7.3)_2$ was still found. These are the principal mathematical features of Boole's *MAL*, deeply influenced by his work on differential equations.

Philosophically speaking, as the references to "mental acts" shows, Boole espoused a kind of psychologism. This word is much over-used in this context, since most logics affirm some relationship between the mind and its linguistic and logical products. So his sense needs specification.

One form of psychologism sought to reduce all knowledge to actual mental processes; Boole deliberately avoided this form (for example, *MAL*, 16). At the opposite pole is a social psychologism, which ignored the inner workings of the mind and relied upon empirical evidence as philosophical basis. It was best represented at that time by John Stuart Mill (1806-1873), whose *A system of logic* was first published in 1843, four years before *MAL*. Much of Mill's book would today be regarded as philosophy of science; on logic he asserted empiricism to the length of claiming even that basic laws such as that of excluded middle were formed by inductions from observation (Richards *1980a*). Boole cited Mill only once in print, on a neutral matter, in the preface of *MAL*; but he objected to Mill's philosophy in a letter of 1855 (page 200), and avoided this kind of psychologism also.

Boole was psychologistic in that he saw his study as a normative account of the correct use of thinking — or, better, of thoughts, the products of thinking, and without concern over the manner of manufacture. He always stuck to this position. In *MAL* "That which renders Logic possible, is the existence in our minds of general notions", and their "elementary laws upon the existence of which, and upon their capability of exact symbolical expression, the method of the following Essay is founded" (pp. 5-6); *LT* opened with a declaration of intent "to investigate the fundamental laws of those operations of the mind by which reasoning is performed"; while in an essay destined for PL "by the term Logic in its primary and most general sense we understand the Philosophy of the Laws of Thought as expressed" (page 126). However, the means of *expressing* thoughts was to be a major issue throughout his logical career.

§8 Between books, 1848–1854

Boole sent out copies of *MAL* to a few friends and colleagues. One of the first to respond was fellow algebraist Cayley, who had already been in touch over developing Boole's idea of an invariant, thanked him in December 1847 for the copy, noting the lack of a cancellation law (7.2) and wondering whether "Has $\frac{1}{2}x$ any meaning" (page 191). In the ensuing exchange Boole stressed his important notion of (non)-interpretability, making analogies between

$$\tfrac{x}{2} + \tfrac{x}{2} = x \text{ and } \sqrt{-1} \times \sqrt{-1} = -1 \tag{8.1}$$

where the combination of two uninterpretables yield an interpretable quantity; Cayley was not impressed.

In the preface to *LT* Boole was to state that he had written *MAL* quickly and "within a few weeks after its idea had been conceived"; in a paper of 1851 he had explicitly regretted its publication (*1851a*, 251). At some time he annotated two copies with additions and comments (Boole *1952a*, 119-124; Smith *1983a*), and he began to develop his theory in manuscripts. In 1848 he wrote one on "The nature of logic" which launches our Part A as Chapter I. Basing his task on the assumption that "The mind is not a passive recipient of the impressions of external things but it has the power of modifying its own conceptions according to the laws of its own nature" (page 3), he expounded upon these laws in more detail than in *MAL*, though largely within the same framework (including on page 10 the same very universal Universe). He verbalised his index law by the case "good good men = good men" (page 7), which he repeated in *LT*, 32 and 43.

Around this time Boole also made his first attempt to write an "Elementary treatise" on his logic; it comprises Chapter II here. He stayed around syllogistic logic, and did not use his index law. However, instead of stressing its forms, he asserted instead that "in general we reason by signs. Words are the signs most usually employed for this purpose" (page 14). These words mark a conversion to the sign tradition in two respects: that signs are primary, and that "language affords the *signs* by which these operations of the mind are expressed and communicated", as Whately put it (*1826a*, 55). It marked a significant change of emphasis from Boole's neutral remark in *MAL* that "The theory of Logic is thus intimately connected with that of Language" (p. 5, typically Endowed with Victorian Capital Letters). Boole's adoption of this tradition was permanent, although he gave the mind a more active role than was usual among its adherents.

In consequence of Boole's new position, elective symbols were demoted in favour of classes; but in their turn classes were only one component, albeit a major one, in the supervening notion of language regarded as a "system of signs" (page 11). In a paper on "the calculus of logic" published in May 1848 in *The Cambridge and Dublin mathematical journal* (as the journal had been renamed in 1846), he reviewed the methods of *MAL* and marked this change by giving a greater presence to classes via this enigmatic identity (or equality?) (*1848a*, 126):

$$\text{"}x1 \text{ or } x = \text{the class X"} \tag{8.2}$$

Boole always restricted his use of language to nouns, adjectives and prepositions (page 14); nouns specified classes within the Universe, such as "men" within "humans"; adjectives similarly determined sub-classes, like "good men" within "men". Prepositions expressed the connectives: "except" (−), exclusive "or" (+) and "and" (·). He could have continued this pattern by adding adverbs to his linguistic repertoire, forming, say, sub-sub-classes of "(not) very good men" from the class of "good men". He seems not to have done so, however; maybe he realised that not all adverbs obey the index law: "very very good men" ≠ "very good men".

So one must be wary of modernising Boole's theories too much. The dash of denoting theory on page 16 does not make him a budding Russell, and he never envisioned propositional functions, or quantification in the sense in which we now understand the term. Similarly, his theory of classes was always the traditional one of part and whole; he never envisioned the set theory of Georg Cantor (1845-1918), which was come into mathematics only in the 1870s, and not into logic until a decade after that, with the very different mathematical logic of Giuseppe Peano (1858-1932) (§14).

Another innovation in Boole's "Elementary treatise" was his distinction of "primary" from "secondary" propositions, the latter being "propositions which express a judgement with respect to other propositions" (page 29). The ensuing analysis was to develop in *LT* as a new theory of hypothetical propositions to replace (7.4) (§9).

Chapters I and II contrast in an important way. The first one treated logic as a science by basing it upon signs; but its successor dealt with "Logic as an Art" in "exhibit[ing] the most useful general forms in which valid argument may be expressed (page 13). The distinction was often stressed at this time; in particular, Whately started his *Elements of logic* with it. Boole was primarily concerned in his manuscripts with its scientific status.

Boole treated much of the subject matter of these two Chapters also in a collection of short essays or a suite of paragraphs on various themes, each given a title. Thus some essays, though maybe not all of them, date from this time. He wrote them in a notebook and removed the folios at a later stage. Chapter III contains from them selected "Extracts", as they shall be cited below.

One extract shows Boole relating signs to words in a language containing not only nouns, adjectives and prepositions but also verbs (page 48). Another one marks the début of his concern with the role of integers in his logic (Panteki *1992a*, ch. 8); maybe Cayley's letter had initiated it. Repetition of x to produce $x^2, x^3, \ldots$ leads to integers, similar to their presence in powers of differentiation in the other index law $(6.3)_1$; so do the expressions $2x, 3x, \ldots$. He defended them on grounds of generality, both as constants and as symbols for operations, since raising powers is an operation in mathematics (page 44). He also examined laws of division, and thought about properties of inverses (page 45). He also pointed to differences, especially that the logical Universe 1 satisfied laws not possessed by the arithmetical 1 (page 45); the cause lies, of course, in the different index laws (6.3).

These notes also relate to an essay written in two notebooks which Rhees published in his edition (Boole *1952a*, 141-166) and which we identify as "Sketch". In it Boole initiated his interest in probability theory; we have not included any other such manuscripts (there are in fact few of them), but a letter of 1849 to the astronomer and mathematician Sir John Lubbock (1803-1865) records the birth of his concern (pages 197-200). Thus he may have started the notes in or around that year; and continued the file for a period; the last extract planned out a book in four chapters to succeed *MAL* which somewhat resembles the "Elementary treatise" but also included the primary and secondary propositions (page 46).

A connected novelty in the notes is the (surprising) arrival of time in Boole's logical scene, due to "the repetition of an operation as in successive

Figure 7:
The house in Grenville Place,
Cork where Boole wrote *LT*
(Photograph I. Grattan-Guinness)

periods of duration" (page 44). In "Sketch" he came to think of his elective
"symbols x, y, z, as representing the *times* in which the elementary
propositions to which they refer are true"; thus x represented the proposition
"it rains" by the symbolic proposition "$x = 1$" (*1952a*, 146). One recalls W. R.
Hamilton's use of time in his definition of irrational numbers in *1837a*; but
their lack of contact (§2) suggests lack of influence. This theory hardly seems
to be a philosophical improvement on the theory of hypothetical propositions
in *MAL* (Hailperin *1984a*); nevertheless he was to use it in *LT*, to which we
now turn.

§9　Logic in *The laws of thought*

Boole wrote this book mostly at his residence in 5, Grenville Place, Cork
(Figure 7). He may have finished it by 1852; it was typeset in Dublin during
the next year, and published by MacMillan in Cambridge and Walton and
Maberly in London early in 1854 (Neil *1865a*, 165-166).[11] Boole and a friend

bore the expense (Boole, M. *1931a,* 27);[12] it is unlikely that they recouped the costs, for it sold badly, much less well than *MAL*.[13] Yet it exhibited a major extension of his logic. He seems not to have recognised the full consequences of his own ideas when writing *MAL*, in that syllogistic logic provided the main guide there; by contrast, it was relegated in *LT* to the last of the 15 chapters on logic.

The preceding 14 chapters were taken up with a much more general and detailed account of Boole's logic, both philosophically and mathematically. Although his psychologism (as specified in §7) was still fundamental, with signs as the key notion in his Logic as a Science, language was very prominent: for example, a major task was *"To deduce a consideration of those operations of the mind which are implied in the strict use of language as an instrument of reasoning"* (*LT*, 42).

Among these operations, Boole reduced the index law from $(6.3)_2$ to

$$x^2 = x \qquad \text{with} \qquad x(1-x) = 0 \qquad (9.1)$$

as a principal consequence (pp. 32, 49). The reduction from the n-th to the second power was first stated in "Sketch" without discussion (*1952a,* 141, 149); he now argued that any higher power of x would lead after factorisation to the uninterpretable term $(1+x)$, which could therefore be discarded (p. 50n). The symmetry of $(9.1)_2$ as a function of x and $(1-x)$ led him to name it "the law of duality" (p. 51), and he deployed this important algebraic property in several places later on in the book. One of them was in his new theory of primary and secondary propositions, explained in ch. 4 and applied in chs. 11-12, with "Time" as a major concept. It also underlay his claim that "we can employ the symbol y to represent either 'All Ys' or 'All not-Ys', since the interpretation of the symbol is purely conventional" (p. 232) and allowed the absence of a symbol for "not".

The balance away from elective symbols towards classes continued strongly, especially after a proposal to read 0 as $0y$ and 1 as $1y$ (p. 47). Boole had already defined the Universe more carefully than hitherto as the "assumed or expressed limits within which the subjects of its [the mind's] operation are confined" in a given "discourse" (p. 42), a move heralded in the "Extracts" (page 44).

A major innovation, heralded in "Sketch" (*1952a,* 151-152), were the elimination theorems; for one and two variables respectively,

$$\text{``}f(1)\,f(0) = 0\text{''} \qquad \text{and} \qquad \text{``}f(1,0)\,f(1,1)\,f(0,1)\,f(0,0) = 0\text{''} \qquad (9.2)$$

(pp. 101, 103). Boole's proof of the first one was rather ponderously algebraic; Harley *1871a* was to show that it rested upon the theory of the roots of equations adapted to variables which took only two values.

The expansion theorems (7.5-6) were given a much more important place, and Boole now saw the role of all these theorems as not only solving (sets of) logical equations but also laying down conditions for any solution to occur. For example with three classes, which class(es) z, if any, satisfies given relationships with given classes x and y, and under what conditions? Boole's method rested on taking all possible basic terms xy, $x(1-y)$, ... and using the appropriate version of (9.2) to calculate each of their coefficients c. If $c =1$, then xy was part of z; if $c = 0$, then it was absent from z; if $c = \frac{0}{0}$, then *any* part "vxy" of xy would be part of z, which therefore could not be specified uniquely; and if some other value arose for c (often $\frac{1}{0}$; or 2, or 576.5 for that matter), then not only was xy no part of z but the condition $xy = 0$ was required for any classes z to be found at all. However, exceptional solutions could still escape the clutches of even this version of the method, as with (7.3). Hooley *1966a* gives a method of obtaining all solutions of a linear equation in two-valued variables satisfying (9.2); Boole does not seem to have spotted it.

Chs. 16-21 of *LT* were concerned with probability theory, which we saw born in 1849 in §8; his main innovation was to interpret compound events as Boolean combinations of simple ones (and thereby increase the 'model-theoretic' character of his algebra). He also considered probabilistic inference, and in *estimating* probability values of logical consequences he manipulated inequalities in ways which anticipate linear programming (Hailperin *1986a*). After *LT* he produced important papers in this area; unfortunately none of his contributions made a major impact.

The last chapter, "On the nature of science, and the constitution of the intellect", shows Boole's psychologism well to the fore: "what is remarkable in connexion with the processes of the intellect is the disposition, and the corresponding ability, to ascend from the imperfect representations of sense and the diversities of individual experience, to the perception of general, and it may be of immutable truths" (p. 407). Mathematics was at the heart of this; and psychology was influenced because the "mathematical laws of reasoning are properly speaking, the laws of *right* reasoning only" (p. 408). Among those laws, index and duality (9.1) were prime, with a long pedigree in ancient Greek philosophy (pp. 411-416); they also deployed "a scientific connexion between the conceptions of unity in Number, and the universe in Logic" (p. 411). Boole now tried to develop the philosophy of his logic within and around such considerations.

§10 Logic after *The laws of thought*

We treat now the manuscripts presented in Part B and the pertinent items from Part D, all written seemingly after *LT* but not destined for the third book, PL (§11). They are to be taken together with one which Rhees included in his edition under his title "Logic and Reasoning" (Boole *1952a*, 211-229).

Science dominated Art in most of the manuscripts in this Part—
explicitly so in the opening note of a collection forming Chapter VI (page
105). Boole continued there with a summary of a philosophical analysis of
logic drawn largely from the sign tradition (he mentioned Locke on page
106); Whately had summarised it as a prologue to his account of syllogistic
logic (*1826a*, 54-56). Boole presented it in more detail in a long essay of 1856
on "the foundations of the mathematical theory of logic", presented here as
Chapter V.[14] In his version of the analysis he paid rather better attention
than usual to the difference between a mental act and its product. The act of
Conception by the mind produced a Concept, such as "men" (or the class of
them); then the act of Judgement of the co-presence of Concepts produced a
Proposition, such as "this is a wise man"; finally, the act of Reasoning
produced a Conclusion, perhaps "wise men are ...", as delivered by his
algebra of logic (pages 88-95).

Throughout this essay, and indeed in all manuscripts assigned to this
and the next Parts, Boole remained in his linguistic territory of nouns,
adjectives and prepositions. To characterise mental actions he set aside
elective symbols (and also time, one may be relieved to note) and envisioned
logic as "conversant with things only as they fall under the general notion of
Class" (page 67). Two pages later he even associated this commitment with
the old and important notion of "second intentions" (or second-order
predicates), which in his time still was ignored; he went back for it to an old
source, Suarez *1605a*. He also considered the intensional and extensional
conceptions of classes (pages 74-76).

Boole rehearsed his basic "Laws of conception" in class terms, imposing
his restrictions on addition and subtraction without discussion (pages 77 and
78 respectively). At a later point (page 92) he claimed a *proof* of the index law
from the argument

$$(x+y)^2 = x+y; \qquad \therefore xy+yx = 0; \qquad \therefore xy = 0 \text{ as solution.} \qquad (10.1)$$

However, as in (7.3), our logician has not sorted out his necessary from
his sufficient conditions. He also deployed and discussed his expansion
theorems and their enigmatic coefficients (pages 98-100, 187-188).

Alongwith his new allegiance to the sign tradition, Boole recognised the
importance of syllogistic logic. In particular, in the essay of 1856 (page 72) he
paraphrased from some edition of Whately that "since Logic is wholly
concerned in the use of language, it follows that a Syllogism (which is an
argument stated in a regular logical form) must be an 'argument so
expressed, that the conclusiveness of it is manifest from *the mere form of the
expression i.e.* without considering the *meaning of the terms*'" (*1826a*, 88).[15]

In all these manuscripts Boole's type of psychologism remained in place.
He developed one aspect in an undated note contrasting belief with
understanding (Chapter XII), and planned out a two-page piece on the place

of freedom within his stern Science of Logic and the laws that the mind should obey (Chapter XIII).

§11 Plans and texts for "The philosophy of logic"

Despite the poor sales of *LT*, MacMillan's seem to have been ready to publish a sequel on "The Philosophy of Logic", and to have stated it to be "nearly ready" in 1857 (Neil *1865a*, 172). But "the announcement was premature", Boole told De Morgan in March 1859 (Smith *1982a*, 77),

> I have written at different times as much as would make two or three books but when returning to a subject I can seldom make much use of old materials. They have lost their freshness & I can only begin again *ab novo*. And that is what I am doing — but — with a modest plan before me, having certain things to say & only desiring to say them. I am not going to set aside anything in the Laws of Thought — but only to interpret now within the province of pure Logic what is done there. When this is done I shall quit for ever.

But by then Boole must have been deep into his textbooks; in the remaining five years of his life, his other publications were several papers in the same area, and his final months were devoted to a second edition of the book on differential equations. It seems that he had stopped working on PL, perhaps in the early 1860s; he implied as much to De Morgan in a letter of November 1862 (Smith *1982a*, 102). On the other hand he had definitely started; his wife had even helped by transcribing some manuscripts in the preparation for the press. There is at least a preface and a table of contents for a four-chapter work (Chapters VIII and IX); we place here also notes that may have been intended for a revised preface, and a long manuscript which looks like an attempt at the first two chapters (Chapters X and XI). The latter manuscript contains comments on De Morgan which suggests that he may have read De Morgan's paper *1860a*; if so, it is among his latest writings on logic. No other manuscript seems definitely to belong to PL.

In his preface Boole stressed again that Logic as a Science was his concern (page 119). In the two drafted chapters he asserted that "the [C?]onception of class really occupies a position of priority in the intellectual order", but that its "narrower Logic of Class" would be developed "with constant reference to that higher Logic [...] as expressed by signs" (page 129); their laws "are a visible expression of the formal laws of thought" such as "the conception of any collection of things as a *whole*" (page 131). Further, in more detail than anywhere else he outlined and then criticised the traditional "Methods of Logic" (page 132) centered on syllogistic logic, both traditionally Aristotelian and as practised in his time by Sir William

Hamilton and by De Morgan (pages 133-141). His main criticisms of syllogistic logic were the emphasis on forms instead of signs, and words to express them, and its limited scope: he gave as an example its inability to state that equilateral triangles were identical with equiangular ones (page 136, perhaps to reinforce Whately's remarks on this example in *1826a*, 42).

De Morgan's contributions to logic are summarised in §13; a word on Hamilton's would be appropriate here. As was noted in §5, he popularised quantification of the predicate, and is often falsely credited as its inventor. In fact his contributions to logic and philosophy were slight, but he helped to raise interest in syllogistic logic, to popularise commonsense philosophy, and to bring German philosophy into Britain. In Boole's terms, he also saw Logic primarily as a Science, and took as basic the laws of identity, contradiction and excluded middle. Boole also agreed with him in construing Logic as an "unexclusive reflex of thought" (page 140), that is, an all-encompassing department of knowledge; he implied that he saw his own logic as more effective for the task. This all-embracing view of logic as Logic is notably different from modest modern approaches, where some logic among the hundreds now available is examined for its local (meta-)properties.

But Boole was quite at odds (at *MAL*, 11-12) with Hamilton's detestation of mathematics (Olson *1975a*, esp. pp. 62-71).[16] Indeed, his foundations of Logic had to differ from those of all his contemporaries because his logic was applied mathematics, and so needed a mathematical basis. This survey of the selected manuscripts ends with his handling of that aspect of his philosophy.

§12 Where was Boole's philosophy of mathematics?

> As algebra is an instrument of thought, applicable to our reasoning upon number and quantity, as its validity for this purpose depends upon the existence of laws such as [distributivity and commutativity ...], so is language an instrument of thought ...
>
> Boole *1848d*

Boole's most extended manuscript on the philosophy of mathematics is rendered here as Chapter XV. Undated, it seems to be part of a book (though not PL), for it contains a "ch. 2" on geometry. He dwelt mainly upon Euclidean geometry, and suggested that our understanding of the different levels of generality of the assumptions in Euclid's *Elements* was not held by its author (page 177). His claim that "it becomes possible to present the entire procedure of the Science [of Geometry] in the forms of consecutive Syllogisms" (page 169) fitted poorly with his allegiance to the sign tradition. It also suggests that he had not read De Morgan's pioneering and far more sophisticated examination of the logic of Euclid's geometry, although it had been reprinted in De Morgan's book *Formal logic* (*1847a,* ch. 1), which was published on the same day as *MAL* and of which he had received a copy. It may have been stimulated by a long discussion earlier in the manuscript of the role of definitions in mathematics; he saw them as "not description but distinction", between those things satisfying a definition and those not (page 172) — like the law of excluded middle, perhaps?

One issue in the relationship between logic and mathematics which Boole seems not to have considered is the danger of a consistency loop. Suppose that one wishes to found (a) logic L on a mathematical theory M. To be worth using, M should be consistent; but that is already a logical notion, no? Today we would instinctively switch into metalogic to handle this question, but that distinction was not properly appreciated until the 1920s. Curiously, in the course of analysing the general linear form (7.7) in *MAL*, Boole had brought up the possibility of inconsistency in the case of "1 = 0, which would indicate the non-existence of the logical Universe" (p. 65); but he did not pick up the point again. However, it lies at the heart of his task.

This was not the only lacuna in Boole's philosophy of mathematical theories. All of his points were necessary but not sufficient for the needs of his algebra of logic. Near the end of *LT* he had asserted, quite legitimately within his philosophy, "The truth that the ultimate laws of thought are mathematical in their form" (p. 407), and in his manuscripts he gave examples; for example from a fragment, that the expansion theorems came only from the basic laws, "not from a consideration of the possible relations of things to each other" (page 187). In another fragment "on reducing the Science of Logic in its technical development under the dominion of mathematical forms" (page 188), he noted the expanding realm of "the Science of Mathematics", which should be "no longer defined in its essential character by the nature of its subject matter but by the forms and method of its procedure" (page 189).

However, Boole was unable to find a philosophy of mathematics rich enough to ground his logic; and it may have been realising this, at least intuitively, which led him to stop working on PL. He needed something like a universal calculus of symbols within which both mathematics and logic could be sited (Laita *1977a*); but he never found it, and in any case the consistency loop might not be broken. He might have found something in the semiotic tradition in France at the turn of the centuries, in which the Abbé Condillac (1714-1780) was father-figure to a group of *idéologues* who emphasised the place of signs on knowledge and thought,[17] and also regarded (common) algebra as the language *par excellence* (Panteki *1992a*, ch. 1). However, like his contemporaries, Boole did not pick up this tradition, although its own parentage goes back to Locke (Auroux *1981a*).

During the late 1850s Boole was spending the majority of his research time on preparing his textbooks on the differential and difference equations; but its subject matter was bound closely to his philosophy of logic by the common factors of signs, operations and language. In the early "Extracts" he gave an example from a method of evaluating a type of multiple definite integrals due to J. P. G. Lejeune-Dirichlet (1805-1859), to which he was alerted by Cayley (Boole *1848b*, 140). In this method the integrand was multiplied "by a factor of which the value is equal to unity in the range which the integrations must cover, and which vanishes outside of this range" (Dirichlet *1839a*, 375); Boole saw this assignment of 1 or 0 as a "good example of the real nature of an elective symbol" (page 45).

A more substantial example is singular solutions of differential equations, which Boole was recalled as teaching at Cork "not like a professor writing a demonstration on a blackboard, but like an artist painting from a vision" (Boole M. *1931a*, 76). Why? As he wrote in his textbook on differential equations (*1859a*, 140), for "the two marks, positive and negative, by the union of which [they] are characterised": solving the equation but lying outside its general solution — a perfect image of $x+(1-x) = 1$. During his last summer he wrote to his wife that his work in the Library of the Royal Society had brought the theory "in a state of unity" (Boole, M. *1931a*, 444), and he studied not only recent work but also the history of the topic (Boole *1865a*, 9-37).

Elsewhere in that textbook occurred an even closer link (*1859a*, 398-399). The example in hand was $f(\frac{d}{dx})$, but this passage could have been in PL:[18]

> In any system in which thought is expressed by symbols, the laws of combination of the symbols are determined from the study of the corresponding operations in thought. But it may be that the latter are subject to *conditions of possibility* as well as to laws *when possible*. And thus it may be that two systems of symbols, differing in interpretation, may agree as to their formal laws whenever they both express operations possible in thought, while at the same time there may exist combinations which really represent thought in the one, but do not in the other. [...]

> Now all special instances point to the conclusion that this is permissible, and seem to indicate, as a general principle, that the mere processes of symbolical reasoning are independent of the conditions of their interpretation. In the few instances we may have occasion to employ, verification will be easy. We take occasion to notice that whatever view may be taken of this principle, whether it be contemplated as belonging to the realm of *a priori* truth, or whether it be regarded as a generalization from experience, it would be an error to regard it as in any peculiar sense a mathematical principle. It claims a place among the *general* relations of Thought and Language.

§13 Outside Boole's universe: Leibniz, Peacock, Babbage and De Morgan

From lacunae to omissions: these four figures are often associated with Boole in some way; yet they appeared but little in his publications of manuscripts. The reasons provide additional perspectives on his work.

Firstly, G. W. Leibniz (1646-1716) arises not only for his dream of a universal language including logic but especially for a passage (transcribed in the textual notes on page 219) in a manuscript *1681a* in which he expressed a law similar to the index law $(9.1)_1$: in those notations, it reads $xy = x$, with x and y as terms and the product read as conjunction. Moreover, it had been published, for the first time, in J. E. Erdmann's edition (Leibniz *1840a*) of philosophical writings; so it was available to Boole (Peckhaus *1996a*). However, it did not exercise influence upon him. Apart from the possibility that he had already envisioned such laws from considerations of monism (§2), he had not seen the edition, and he was informed of the passage by his friend Robert L. Ellis (1817-1859) only in 1855, a year after *LT* appeared (Harley *1867b*, 5; compare Boole, M. *1931a*, 1132). Boole's first reactions are reproduced here among the fragments (page 185 and the self-confessional page 188).

In any case, Leibniz had not developed his remark into any *theory*, so that the anticipation is not substantial. In any case, which theory? When Peano began to develop his mathematical logic in the 1880s, he also cited Leibniz as father-figure ...

Previously Boole had not mentioned Leibniz much; a few places in passing (for example, pages 133 and 179 here), and on the law $(7.1)_1$ of contradiction from a Leibniz publication (*LT*, 240: in Chapter XIV he noted this law in Aristotle). Despite Erdmann, the logical and philosophical writings of Leibniz were far less well known then than they are to us, in the wake of Couturat *1903a* and later research. Indeed, Leibniz was still regarded in Britain as the stealer of Newton's calculus, although De Morgan had recently started to undermine that story with proper historical research (Rice *1996a*).

Our next two figures were founders of the "Analytical Society" at Cambridge (1812-1813), who played an important role in the revival of English mathematics by adopting Continental methods. George Peacock (1791-1858) later wrote a large "Report on the Recent Progress and Actual State of Certain Branches of Analysis" (*1834a*), which historians often cite as a landmark text in setting the stage for the development of English algebras. His emphasis on symbolic methods and the (non-) interrelativity of a formula in a given algebra doubtless reinforced awareness of the issues raised by the French mathematicians cited in §6. His "principle of the permanence of equivalent forms" asserted that if an equation A = B held in an algebra for specific conditions on the components of the expressions A and B, then it held in general; and Boole's analogies such as (6.1-2) certainly provided example. But *citation* of Peacock by the English is less common than may be expected; in particular, Boole seems not to have mentioned him at all. This silence is in fact somewhat surprising; for Peacock had emphasised the role of signs in algebra, and his line of thought shows some kinship with Locke (Durand *1990a*), and so with the sign tradition.

The third figure is Peacock's Analytical colleague Charles Babbage (1791-1871). He is sometimes mentioned with Boole; for his long concern with calculating machines, which ran on throughout Boole's youth and adulthood, seems so akin to our use of Boolean algebra. However, the two men had no particular congress, and no manuscripts of one seems to dwell on the other. On Boole's side, as we have seen, in both logic and for educational practise he thought that the mind was capable of original action, such as grasping general laws from particular cases (*LT*, 4; Boole, M. *1931a*, 5-6), and so he would not have welcomed the association of his logic with the *repetitive* actions of computing; whether Babbagean mechanical or modern electrical. For his part, Babbage disliked Boole's practise of allowing the algebra of differential operators to be interpreted as operations and as objects.

This contrast is stronger than it may seem; for in mathematics Boole and Babbage had an important common factor. In the 1810s young Babbage had worked on another new algebra encouraged by Lagrange's foundation of the calculus (§6) — functional equations, in which functions rather than unknown constants were sought. For example, find the functions f (if there are any) for which

$$f(x+y) = f(x)f(y) \text{ for all real values of } x \text{ and } y. \tag{13.1}$$

The theory attending this algebra contained various features which are evident also in his work on computing; iterations, algorithms, and especially signs, whether physical, mathematical or mental (Grattan-Guinness *1992a*).

The first systematic account of functional equations was written by our fourth figure in this section: De Morgan, in a long article *1836a* for the *Encyclopaedia metropolitana*. It must have influenced his most original contribution to logic, in 1860, a syllogistic logic of two-place relations (Merrill *1990a*), for the analogy between this logic and functional equations is strong; inverses, compounding, and so on. Boole used functional equations occasionally, such as when finding his general expansion theorems (7.5-6) (*MAL*, 60-63), but he never noticed this important aspect of logic. Nor did he seem to react to De Morgan's paper, although by 1860 his main interest lay in differential equations.

Previously De Morgan had extended considerably the realm of syllogistic logic by applying various algebraic techniques, of both theory and signs (Panteki *1992a*, ch. 6): the quantification of the predicate (§5) is an important example. (De Morgan *1966a* is an edition of his main papers.) An algebraist by mathematical inclination, he and Boole were the most significant algebraic logicians of their time, and each man was influenced by a new algebra. But their two approaches were quite different; although they corresponded regularly (Smith *1982a*), neither used the other's logic, and they communicated more than discussed (Corcoran *1986a*). Thus Boole's comments on pages 138-141 noted in §11, his most extended statement on De Morgan's logic, were not sympathetic.

§14 Boole's immediate legacy

While always staying with his psychologistic view of logic, in the sense explained at the end of §7, Boole made various changes while seeking its foundations. The most significant one was away from syllogistic logic to the sign tradition; language gained ground along the way, and so did classes, at the cost of elective symbols. Time made an appearance, though seemingly for a short duration. The algebra received a very original (although not complete) development; its central place in the logic required a rich mathematical basis, which he was not able to find.

Boole's contemporaries knew little or nothing of his manuscript ruminations. After his death, his reputation rested largely on his textbooks in differential equations, which continued to be well used after his death and indeed are still in print. But his differential operator methods became generally eclipsed by other techniques from the 1870s, and his work on logic remained rather marginal (Panteki *1992a*, ch. 9).

Boole's most important initial follower in logic was Stanley Jevons (1835-1872); they corresponded in 1863 and 1864, just before his death. The main topic was considerable disagreement over "+", using "$x+x$" as a test case (Grattan-Guinness *1991a*). Jevons had no time for Boole's restriction, which rendered the expression uninterpretable[19]; in his book *Pure logic* (1864), he wanted to allow that

$$x+x = x. \tag{14.1}$$

Shortly after Boole's death the chemist Benjamin Collins Brodie (1817-1880) tried, unsuccessfully, to develop a similar algebra for chemical combination (Brock *1967a*). John Venn (1834-1923) was the sole close adherent in Britain, mainly in his textbook *Symbolic logic* (1881, 1894); to make the theory more easy to grasp he introduced diagrammatic representations of syllogisms — not the ones which are misnamed after him, which are due already to Leonhard Euler, but a rather clumsy procedure in which all possible overlaps of convex shapes corresponding to the given predicates are shown and the non-empty ones in a given set of premises are examined.[20]

Among other British semi-followers, Hugh MacColl (1837-1909) made an innovation that was to be so durable that Boole is sometimes credited with it: that "The calculus of equivalent statements", or the propositional calculus, could be construed as a Boolean algebra, with addition and multiplication read respectively as disjunction and conjunction, and equality as equivalence (MacColl *1877a*). This proposal, and the differences between Boole and Jevons, were among the topics reviewed in a lengthy and influential survey *1892a* of "The logical calculus" made in the philosophical journal *Mind* by the Cambridge logician W. E. Johnson (1858-1931). A few years earlier Mrs Sophie Bryant *1888a* had swung against the trend in its pages, claiming that Boole's logic should be understood as an operational calculus and lamenting the preferred reading in terms of classes.

Outside Britain, Boole's logic gained some publicity during the 1870s. The Frenchman Louis Liard (1846-1917) wrote an article *1877a* on it, and followed up with a book *1878a* "On the contemporary English logicians" which became quite popular although logic was not in general a well received topic in France. G. B. Halsted (1853-1922) performed a similar service in the USA with a survey article *1878a* (and also a piece *1878b* in *Mind* on Jevons's modifications).

The most significant advance began also in the 1870s: the fusion of Boole's algebra of logic with De Morgan's logic of relations. This was effected by Halsted's compatriot C. S. Peirce (1839-1914), who also modified addition to (14.1), independently of Jevons. The Scottish mathematician Alexander MacFarlane (1851-1913) made a modest contribution of this kind in a short paper *1879a* (and in a contemporary book *1879b* he modified Boole's logic by distinguishing nouns from adjectives).

In Germany the German philosopher Hermann Ulrici (1806-1884) gave Boole's logic publicity, starting out from a long review of *LT* in 1855 (Ulrici *1855a*). It also came to bear upon a massive enterprise of algebraic logic undertaken by Ernst Schröder (1841-1902); but he drew more directly on the algebra of Hermann Grassmann (1809-1877) and the logic of brother Robert Grassmann (1815-1901), and especially on Peirce. This work was taken up in turn. In Russia from the mid 1880s, principally with P. S. Poretskii (1846-1907) (Styazhkin *1969a*, 216-253).

However, from the beginning of this century the whole algebraic tradition in logic began to be eclipsed by the development of mathematical logic with Peano and his school and then with the logicism of Russell and Whitehead. It drew on completely different principles, including Cantorian set theory, and furnished mathematical quantities such as real numbers and lengths as well as qualities such as true and false; indeed, logicism was logic applied to mathematics instead of Boole's vice versa (Grattan-Guinness *1988a*). The failure to produce an edition of Boole's manuscripts in the 1900s (§3) was a sign of those times in symbolic logic.

The psychological and religious aspects of Boole's logic largely disappeared with him, and they do not feature in his manuscripts. Perhaps he could not see how to build them in effectively — and since they came from his Dissenting heart, how to convey them honestly. But there seems no doubt that they always lay close to the centre of his logic. Boole's Boolean algebra was another new algebra in the time of their proliferation, but one set apart from the others not only by the index law $(6.3)_2$ but especially consequences such as (7.1) involving the all-important Universe 1. This remained with him from youth throughout his mature career, especially in the image of the unique Godhead and the ecumenism encaptured by Frederick Denison Maurice. As he lay dying, in only his 50th year, he had Maurice's portrait set by his bedside (p. 48):

> I said, "Are you still willing to leave the future entirely to God?" He nodded, and there came over him a smile of such wonderful peace that I would not speak again; and so he lay quiet until the end.

Part 2: Boole's Psychologism as a Reception Problem

Gérard Bornet

§15 Introduction

More than 125 years after Boole's death, understanding his work is accompanied by many difficulties of interpretation arising from the differing backgrounds of the reader on the one hand and the author on the other. The greatest obstacle to an adequate reception today is surely the psychologism that pervades all Boole's logical work (§§7, 9, 10, 14). In consequence we have sought to prevent an over-hasty dismissal of Boole's position by the reader and to encourage a close look at what Boole is actually asking at his time rather than a simple change of subject.

In this second part we briefly discuss three aspects in which reception of Boole is hampered in connection with the word "psychologism". First (§16) the epistemological framework in which the German mathematician and logician Gottlob Frege (1848-1925) poses the problem of psychologism. Then (§17) we show that Boole must be exonerated from Frege's criticism. The second aspect deals with the question of art versus science. Here, the fact that in the course of time the centre of general interest shifted from logic as science to logic as art (§18) causes a reception problem. Doing Boole justice means considering his psychologism as part of logic as science (§19).

The problems touched on so far relate to the change in the understanding of logic in general and may crop up in connection with other thinkers of this era. But the third aspect is essentially Boole's own peculiarity and addresses difficulties faced even by his contemporaries. In this connection today's attitude is to see these features of Boole's system as quirky and therefore to be neglected. It will first be pointed out (§20) that Boole does not use the psychological terms of ordinary language in their full verbal meanings (compare also page 76). Then (§21) the existence of a whole group of signs central to Boole's system must be emphasised; signs which can have no significance at all in the psychologistic sense because they are "uninterpretable" in accordance with definition.

§16 Frege's psychologism criticism

Gottlob Frege, Professor of Mathematics at the Jena University 1879–1918, is regarded as the founder of modern logic and philosophy of language. His work influenced thinkers like Ludwig Wittgenstein (1889–1951) and Rudolf Carnap (1891–1970) so deeply that he is the so-called father of today's analytical philosophy. A characteristic of his teaching included a strict anti-psychologism (of which more later) and so an avoidance of anything that awoke the merest suspicion of psychologism became a trait of his following. Thus among the ranks of the most likely candidates for interest in Boole as a philosopher, the use of psychological terms in mathematics and logic were no longer seen as *de rigueur*.

However, the sweeping verdict is at times applied to variations not actually included in Frege's criticism, among them, as we shall show, those of Boole as the author of *Laws of Thought*. For example Bertrand Russell ludicrously suggested that Boole had given *LT* the wrong title: "He [Boole] was ... mistaken [!] in supposing that he was dealing with the laws of thought: the question how people actually think was quite irrelevant to him ..." (*1901a*, 366). Many later authors who, like Russell, mentioned Boole only in passing, had, with him, never read Boole's major work,[21] having been put off because the book's title seemed to signal contents irrelevant to logic, namely the question of "how people *actually* think".[22] But it was specifically to avoid misunderstanding in this respect that it was declared, in §7 above, that Boole wanted to give "a *normative* account of the *correct* use of thinking" (our italics).

The false reading of the title testifies to the triumphal march of experimental psychology, which barely existed in Boole's time. The first institute in this field was established for Wilhelm Wundt (1832–1920) in Leipzig in 1875. Up to that time psychology and philosophy had in common that they typically tackled problems theoretically. And as late as the early 20th century, it was not unusual for one university chair to cover both disciplines. In his criticism during a period of radical change, Frege was basically attempting to claim a separate field of study for Logic. Thus he pressed in *1918–19a* (implicitly already in *1884a*) for an autonomous *third dominion*. In this way he could make logic independent of nature on the one hand and individual souls on the other, thereby freeing himself from the authority of Physics and Psychology. Interestingly, a similar move by Boole is found in a fragment of the Nachlass (MS B99.7):

> The phenomenal study of things belongs to physics or to psychology, the inquiry into their absolute nature if indeed such an inquiry be possible is the business of metaphysics. Logic has other objects and is concerned with other relations – but as the relations with which it is concerned are universal (for all existing things can be contemplated under the notion of class or kind) Logic stands related to all other sciences.

As is well known, Frege (particularly in *1884a*) put forward two arguments against a psychologistic foundation for mathematics: that it robbed mathematical reasoning of (a) its *strict truth* because it must be inductively gained, and (b) its *general validity* (in the sense of intersubjectivity) because it would in the end relate to subjective facts. His arguments are so fundamental that they disprove the possibility of a psychologistic foundation for any science – and in this sense they can be applied to Logic too. By insisting on strict truth Frege excludes the use of experimental methods in mathematics; and by stressing the point of general validity Frege makes it obvious that the advocate of psychologism is left with nothing he can objectively speak about.

Consequently only those varieties of psychologism fell under his verdict which (a) used inductive methods or (b) related to subjectivity, that is affections of individual human souls. In no way did Frege intimate that all dealings with thought necessarily drew his criticism. On the contrary, in his *Begriffsschrift* he himself proposed a formula language for pure thought, offering the best counter-argument. It remains to be added that despite the fact that Frege did indeed criticise Boole (§18), he did not reproach him for psychologism. Knowing how quarrelsome the Professor of Jena was, that equals exoneration.

§17 Boole's anti-inductivism and the problem of introspection

In fact neither of Frege's anti-psychologistic arguments applies to Boole though we can here only conclusively prove this for the first, the one from strict truth. For the second, the one from general validity, the reader is referred to page 68 or 1952a, 226, where Boole anchors the laws of logic in the intelligible field, which is per se intersubjective. A psychologistic reading of Boole in the line of the second argument is further also invalidated by the circumstance discussed in §21 that Boole admitted signs which he expressly described as uninterpretable – by which he meant that their meaning in the psychologistic sense was not accessible to the human mind.

Frege's first argument is quickly demolished for here Boole is unmistakable (*LT*,4; compare also page 32):

> ... the knowledge of the laws of the mind does not require as its basis any extensive collection of observations. The general truth is seen in the particular instance, and it is not confirmed by the repetition of instances. [...] a general truth in Logic ... is made manifest in all its generality by reflection upon a single instance of its application. And this is both an evidence that the particular principle or formula in question is founded upon some general law or laws of the mind, and an illustration of the doctrine that the perception of such general truths is not derived from an induction from many instances, but is involved in the clear apprehension of a single instance. In connexion with this truth is seen the not less important one that our knowledge of the laws upon which the science of the intellectual powers rests, whatever may be its extent or its deficiency, is not probable knowledge. For we not only see in the particular example the general truth, but we see it also as a certain truth,—a truth, our confidence in which will not continue to increase with increasing experience of its practical verifications.

Naming no names (cf §7) this was certainly aimed at Mill, just the person Frege (*1884a*) explicitly targeted with his first argument. In this instance Boole the seeker of the laws of thought is at one with the creator of a language for pure thought.

However, there are also differences in their positions. Among other things, Frege designed his *Begriffsschrift* for the purpose of being able to do away with intuition in proof. He succeeded in this by making the unbroken chain of evidence demonstrable by signs and rigorous detailed proofs. Boole on the other hand, in the intellectual field too, speaks of "observation" and "perception" as in the quotation above (*LT*, 4). Today's reader will, however, not be content with the then-current usage of terminology and will require a more exact picture of the mental processes described. Plain from his presentation is in fact only that a "usual" observation cannot be meant, as otherwise it would remain incomprehensible why a repeated examination should not yield a sharper picture.

On the problem of introspection as here addressed, little information is available in Boole. Most interesting is section 9 of the introductory part of chapter iv (pages 52-53) where "facts of external observation" are opposed to "facts of consciousness". In this connection an "appeal to consciousness" or a reference to "the direct testimony of consciousness" appears as a matter of course. But Fragment 1 on page 179 reveals that Boole was quite aware of the special importance of the ability of introspection. If we find him here far away from a satisfactory answer, we must honestly ask ourselves whether one exists at all. For Boole, more closely considered, touches on one of the "eternal" questions of philosophy. Our commentary in §18 points out that progress since Frege, broadly speaking and confined to the matter in hand, lies not in better philosophical answers but simply in the avoidance of the relevant questions.

§18   Art versus Science

> Skill in the processes of art or science applied is one thing. A thorough comprehension of the grounds upon which those processes rest is another.   (Boole, page 124)

The *art* of logic consists in *using* a given symbol system in one sphere. To this, skills in symbol manipulation must also be added, for the greater the skill of working in symbols, the easier is the employment of them. Against that, *science*, in its usual meaning, delivers *reasons* and says *why* something is as it is (compare page 13). That is in every way a philosophical problem, and in Boole's time "philosophy" and "science" were nearer together than they are today. Natural science instruments were wont to be described as "philosophical instruments", and Boole himself speaks repeatedly of a "philosophical language", meaning thereby a "scientific language" (e.g. *MAL*, 5). In the epistemic position of Naturalism this close relationship between philosophy and (natural) science is still alive today.

In the contention between art and science, two problems for the reception of Boole present themselves. In the first place, since Frege logic as art has developed, bringing with it a natural tendency for later recipients to regard earlier logicians as "artists" and not as "scientists". And secondly this has been strengthened by Boole's own attempt to dissociate himself from metaphysics (see quotation from MS B99 in §16); this despite the fact that the borders between metaphysics and fundamental science come down to the simple expression of a value judgement. This section deals mainly with the first point, while the second forms the kernel of the next one. However, one remark should be made in advance.

When Boole emphasised that he was not dealing in "the research of causes" (already in *MAL*, 11ff), it was mainly with the purpose of avoiding metaphysical discussion. In this sense he quite openly declares in *LT*, 11 that (in his book) he would not provide an answer to the question *for what reason* "the ultimate laws of Logic are mathematical in their form". And he continues in a distinctly audible undertone: "Such knowledge is, indeed, unnecessary for the ends of science, which properly concerns itself with what is, and seeks not for grounds of preference or reasons of appointment." Taken literally and isolated, this would show that Boole was not a scientist in the sense first indicated above. But that he is in fact such, is, for example, clear from page 141, where we find him searching for *ultimate* laws of thought (emphasis added). For him the difficulty in a "science of logic" lay not in the ascertainment of general truths of logic – these he recognised at a glance – but in the definition of those which are "primary and fundamental" (LT, 4f).[23]

In the public consciousness, however, such statements remained by the way: Boole was above all known as the author of *MAL* (§2), that is to say as an "artist". This fact led to the philosopher Edmund Husserl (1859–1938) being able without quibble to offer him as an example of a person who could be "a splendid logical technician" and at the same time "a very middling philosopher of logic" (*1891a*, 9). This assessment has something tragic about it for Husserl was another such man, who, at the turn of the century, came down most heavily on the side of a logic as science. He would have been highly delighted by Boole's differentiation between a *noeton* (i.e. an intelligible) and an *aistheton* (i.e. a sensible) (page 68 or *1952a*, 226) but the writings in which Boole sought to bring this distinction to fruition had not then be published.

So we see how Boole, in the eyes of later recipients, fell between two stools. By not getting to the grips with metaphysical general truths of logic, he was dismissed as a "middling philosopher" by those interested in such truths. And since logic as art had bounded forward with Frege, he soon seemed a "middling technician" too (in contrast to Husserl). And so Frege discerned, before all else, in Boole "lack of art"[24], as the following quotation (*1880–81a*, 12) testifies to:

> If I understand him aright, Boole wanted to construct a technique for resolving logical problems systematically, similar to the technique of elimination and working out the unknown that algebra teaches. To this end, he represents judgements in the form of equations that he constructs out of letters and arithmetical signs such as +, 0 and 1. ...

In the main these means fulfil their purpose, at least as far as the range of problems that Boole has in mind are concerned. But one may think of logical problems lying outside this range.

Frege further found fault in that Boole's system for mathematical reasoning was "completely unsuited" if only for the fact that "[Boolean logic] employs the signs +, 0, 1 in a sense which diverges from the arithmetical one. It would lead to great inconvenience if the same signs were to occur in one formula with different meanings." (1880a, 13). As right as Frege is here, his objections refer to Boole's *art* of logic alone and not to his *science*, which is not mentioned at all.

§19 Psychology as science

So much is plain: psychological considerations get logic as art no further. We shall see in §21 that Boole of all people was aware of this. But in the realm of logic as science the role of psychology is different. Here it provides not skills but the necessary answer to the question of the subject area covered by the science being referred to. Insofar as this question is thought admissable, it can, even today, only be satisfied with a psychologistical response.[25] For as Boole correctly remarked the term "logic" comes from the Greek word λόγος, where it "signifies not only the inward thought but also its outward form or manifestation" (page 126).

However, the relationship between art and science has changed beyond recognition. Although it is today possible as a logician to devote oneself to art for art's sake and, say, concentrate exclusively on the syntactic properties of a calculus, in the first half of the 19th century such an attitude was absolutely unthinkable. Joan L. Richards explains why English mathematicians did not develop an abstract algebra as understood today despite possessing all the elements for it. Because such an algebra was not a goal worthy of endeavour (1980a, 345, emphasis added):

> The issue which separates early English algebraic development from modern algebra is ultimately grounded in a fundamental difference between 19th- and 20th-century views of truth. For an English mathematician in the early 19th century to devote himself to developing an abstract algebraic system would have required a major change not only in the interpretation of the nature of mathematics, but more generally in the view of human knowledge. ... In the 19th-century English view, mathematical investigation involved the scientific study of an *external subject matter*, and not merely the logical development of theorems from given axioms and definitions. The truth of a mathematical investigation, the goal after which the mathematical researcher was striving, lay in the *subject*

matter his axioms and theorems *described*. The empty form of the mathematical structure had no truth or value of its own.

Without external subject matter, no science. On this and no other grounds, W. R. Hamilton (*1853a*), among others, felt obliged to develop algebra as "the science of pure time" (§8). It was not his interest in time that swayed him, but the status of algebra as science (also see Øhrstrøm *1985a*). A proper assessment of Boolean psychologism must take this status into account as well.

Anyone wishing to take Boole's demand to be seen as a scientist seriously must also take note of a premise widespread (though not undisputed) at this time: the possibility of a science being strict. This pre-supposes the means of statements which are necessarily valid and which despite being based on experience cannot be refuted by this. Such statements cannot be inductively gained (as we observed already in §16 in connection with Frege). If such a supposition seems rather strange to us today, it must be remembered that only sixty years lie between the publication of Boole's *MAL* and Immanuel Kant's "Copernican Revolution" (*1787a*), in which the latter sought to rescue strict science from scepticism. And so Boole could confidently write in *LT*, 3:

> It is unnecessary to enter here into any argument to prove that the operations of the mind are in a certain real sense subject to laws, and that a science of the mind is therefore *possible*. [...] Let the assumption be granted, that a science of the intellectual powers is possible, and let us for a moment consider how the knowledge of it is to be obtained.

From the acceptance of a possibility of a strict science of the intellectual powers springs more than at first meets the eye. For instance, such a science is inconceivable without a suitable means of expression. It was obvious that Boole, unlike us, would see these means in the language of mathematics, for had this not already made physics a "strict science"? Psychology can be practised as a strict science only if its fundamental laws have a certain simplicity in the manner of Boole's "Laws of Conception" (pages 76ff). Laws of experimental psychology are, in contrast, neither necessarily valid nor simply formulated.

In order to understand the role the mind plays in logic as science for Boole, he must be firmly placed in the "sign tradition" (§5). In this tradition the processing organ for signs is the mind, not the brain. It would never have occurred to Boole to found his logic on the "general laws of nervous actions" as, for example, Peirce did in *1880a*. Such speculation did exist at Boole's time. For instance, Dr. Alfred Smee (1818–1877) proposed an "adaption of process of thought to algebraic formulæ" in *1851a* which is "electro-biologically" grounded. Again, I. P. Hughlings (dates not known) could not resist enriching this introduction to the then-modern logic *1869a* with thoughts on nervous actions. So it is hardly surprising that Boole's psychologism was associated with such thought processes. Yet it struck out in a quite different direction in his philosophy. And thus we come to that trait in Boole's teaching that is most often overlooked or misunderstood.

§20 Boole's structuralism

The fact that Boole did not have the structural terms of modern mathematics at his disposal has already been noted (§19). There is, however, in *philosophy* a more general concept of structure whereby the expression "structure" describes the *relationship* between elements of a composite whole. In the 20th century, structures in linguistics, ethnology and psychology have been examined in this sense. Boole himself did not speak of structures and treated them first and foremost as not having an independent existence. He comes over as a precursor of structuralism, however, when he publicly declares (*LT* 39): "The object of science, properly so called, is the knowledge of laws and relations". Of the entities connected through these laws and relations he intentionally says nothing.[26] A fine example of Boole's "structuralism" can be found on page 157 (the inverted commas indicate a pre-form, not modern coinage), where the author, in a discourse on belief, is not concerned "to investigate the nature of that mental act which we designate belief, but to state some of its general relations."

If science is concerned only with the structure of its subject, an interesting and almost always overlooked consequence arises for the meaning of normal language terms: they are over-defined. Boole illustrates this with the example of controversy regarding cause and effect (*LT*, 39). One person may see nothing deeper in this than a regular succession in time whereas another might insist that there is an innate connection between the perceived phenomena. The two persons may endlessly quarrel over denotation, but even if the one accepted the words at face value and the other sought deeper significance in the nature of causality, they would be agreed on the fact that one of the two phenomena involved followed the other in time. And with this they recognise, according to Boole, "a common element of scientific truth, which is independent of their particular views of the nature of causation." If Boole were to speak of causality – to stay with this example – he would mean with this term only the (uncontroversial) "common element of scientific truth".

This observation is as valid in the sphere of semiotics for the argument between conceptualists and nominalists (page 70f) as in the sphere of psychology, where Boole sets the position of the idealists against that of the sceptics. Here too, there are common threads – and only these interested him as a scientist (*LT*, 40):

Let it even be granted that the mind is but a succession of states of consciousness, a series of fleeting impressions uncaused from without or from within, emerging out of nothing, and returning into nothing again,—the last refinement of the sceptic intellect,—still, as laws of succession, or at least of a past succession, the results to which observation had led would remain true. They would require to be interpreted into a language from whose vocabulary all such terms as cause and effect, operation and subject, substance and attribute, had been banished; but they would still be valid as scientific truths.

Boole is clear that the terms used "in the science of the intellectual powers become, in expression at least, almost necessarily mixed up with the modes of thought and language, which betray a metaphysical origin". (*LT*, 39). Nevertheless, he did not renounce employment of the suspect terms (*LT*, 41):

The course which it appears to me to be expedient, under these circumstances, to adopt, is to avail myself as far as possible of the language of common discourse, without regard to any theory of the nature and powers of the mind which it may be thought to embody. For instance, it is agreeable to common usage to say that we converse with each other by the communication of ideas, or conceptions, such communication being the office of words; and that with reference to any particular ideas or conceptions presented to it, the mind possesses certain powers or faculties by which the mental regard may be fixed upon some ideas, to the exclusion of others, or by which the given conceptions or ideas may, in various ways, be combined together. To those faculties or powers different names, as Attention, Simple Apprehension, Conception or Imagination, Abstraction, &c., have been given,—names which have not only furnished the titles of distinct divisions of the philosophy of the human mind, but passed into the common language of men. Whenever, then, occasion shall occur to use these terms, I shall do so without implying thereby that I accept the theory that the mind possesses such and such powers and faculties as distinct elements of its activity. Nor is it indeed necessary to inquire whether such powers of the understanding have a distinct existence or not. We may merge these different titles under the one generic name of *Operations* of the human mind, define these operations so far as is necessary for the purposes of this work, and then seek to express their ultimate laws.

So Boole argues that there is an ideal language, believing himself to have discovered it in the form of the then-current algebra. His psychologism, however closely it may be defined, cannot be made a case by simply pointing to the notorious woolliness of the psychologistical terms involved. For not only did he himself draw attention to this, he even sought clarity through "translation" into an "ideal language".

§21 Uninterpretable symbols

This last aspect of Boole's psychologism provoked controversy from the very beginning. Soon after *MAL* was published Arthur Cayley raised queries (§8, Chapter XVII). Again in 1863, the subject reared its head in correspondence with Stanley Jevons (§14; Grattan-Guinness *1991a*, 26f.) and later Rudolf Hermann Lotze in *1880a* (quoted in Hesse *1952a*, 76) made insinuations.

What is it all about? Boole applied algebra for the purpose of logical reasoning by forming the premises into algebraic equations, solving them and transcribing the result into a conclusion. That many found the intermediate steps logically uninterpretable did not only not worry him at all (compare pages 43 and 146-148), but indeed was a circumstance to be taken into account, otherwise it would be "manifest, that no such thing as a general method in Logic is possible". (*LT*, 67). At the same time, Boole conceded,

> ... that this apparent failure of correspondency between process and interpretation does not manifest itself in the *ordinary* applications of human reason. For no operations are there performed of which the meaning and the application are not seen; and to most minds it does not suffice that merely formal reasoning should connect their premises and their conclusions; but every step of the connecting train, every mediate result which is established in the course of demonstration, must be intelligible also. And without doubt, this is both an actual condition and an important safeguard, in the reasonings and discourses of common life.

The last sentence of the quotation betrays the source of a fundamental misunderstanding in the assessment of Boole's psychologism. For there is almost an acceptance that anyone who psychologistically interprets premises and conclusions of logical reasoning grounds the *validity* of the final account by making the steps on the way intelligible. Yet for Boole, the validity of the *logical* ending did not depend on given psychological factors but on the laws of symbols (compare page 70).[27] That is why, when he later sought popularisation, he defined logic as "the Philosophy of the Laws of Thought *as expressed*" (page 126, emphasis added), and he began to stress that he was concerned with the *formal* laws of thought.

The comprehensiveness of language transcends our imagination. Boole declares this (page 41) when he says, "it is not true in thought that every operation can be performed upon every subject but it is possible in language to represent any operation as performed upon any subject, the laws which that operation obeys when possible in thought being still observed." In the same sense, Boole prefers the term "notion" to the term "concept" "because the latter seems to indicate the formation of a mental picture or resemblance" (page 68).

This distancing from a psychologistical interpretation of all terms becomes understandable when Boole's work on differential operators in the calculus is recalled (§6). He was also aware that the notion of limit was notoriously difficult to specify (pages 38-40). As a mathematician, Boole was well aware of its limitations and he always maintained this when he began to apply mathematics to logical problems. Thus Boole wrote to Penrose (pages 200-201; compare also page 175):

> "It seems to be a law of human reason that we can in various instances affirm propositions ... respecting things which we can only picture or represent to ourselves as the limits of an indefinite process of abstraction. [...] [T]he object of all pure science are things which it would baffle all our efforts to represent by the power of imagination only and which can only be approved as the attainable limits of thought, but concerning which nevertheless the most vigorous of all propositions can be *affirmed* ...

We have now presented the factors that should be weighed up in an interpretation of Boole's psychologism. The question as to whether these allow for a coherent whole must be left open. Suffice it to say that we feel we have shown Boole to be a great deal more interesting than legend leads one to suppose.

Part 3: Remarks on dating and editing the manuscripts

As was explained at the end of §2, the Royal Society collection has become rather disordered in the course of consultation and transcription over the decades. In recent times some valuable cataloguing has been effected by Panteki in the late 1980s and continued by Bornet in the course of his selection and transcription for this edition; both scholars have placed their unpublished partial guides in the collection, to join one made by Alicia Boole and Falk. We provide readers of this edition with the required information in the Textual Notes, where a general explanation is followed by the details required for each Chapter.

§22 Remarks on dating

Apart from the letters, no manuscripts are dated by Boole; thus chronology is a matter of conjecture. A few papers contain conclusive evidence by direct reference in the text such as "the present year 1848" or "this is written two years after *LT*"; the rest contains only hints. The two most important events regarding a chronology of the material are the publication of *LT* in 1854 and the marriage to Mary Everest the following year.

Mary helped George in the task of transcribing the manuscripts and in preparing them for the press. When she thought about depositing the unpublished works of her late husband she noted on the manuscript if it was known to her and therefore "later than 1855". But this must not necessarily mean that the work originated after her marriage. It could well be that only the transcription was then made.

Regarding *LT*, the only clue for dating used here is an explicit reference in the text to this work. We did not consider the state of the theory presented in the paper when judging if it is prior or later than *LT*. Apart from the difficulties of such an undertaking there was usually better evidence to determine the date of the manuscript or the material was too fragmentary for such a consideration.

§23 Remarks on editing

We aim to retain the original spelling of Boole, including his sparing use of quotation marks. We keep his use of capitalization, although it sometimes looks a little arbitrary. We have rendered his underlinings as *italics*. In some cases he uses a vinculum instead of (brackets). This we also retained regardless of the danger of misunderstanding because in logic a vinculum often marks negation. Obvious mistakes of Boole have been corrected silently. Boole frequently uses "as" where today "just as" or "such as" would be used. Prescriptive modern commas have not been inserted where the meaning is clear.

Some of the manuscripts are in a unfinished state and/or written very hastily. It is not uncommon that references within the text are not supplied so that we only have an empty pair of brackets ( ). Where Boole left a totally empty space for a later reference we complete the brackets silently for uniformity.

If a manuscript is in neat handwriting this is stated, but not otherwise. The transcriber is (with one exception, see chapter II) not known, for possible persons see §3 of the Introduction.

The footnotes of the manuscripts have no explicit reference in the text. Such references are supplied here silently. Footnotes belonging to the text itself are marked "*" and "†". The figures 1, 2, 3, ... in different print from the basic text serve as reference to the Textual Notes at the end of the book.

Page numbers of the original version that appear in square brackets in the text thus "[XX.nn]" concern the marking of the then end-of-page. The "XX" identifies the document, "nn" the page number. Details can be found in the Textual Notes pp. 206ff.

Works mentioned by Boole are identified in the Textual Notes with the exception of his *LT*. In certain passages there would be simply too many references. All Greek expressions and texts appear in the Textual Notes in Latin script and where the meaning is not immediately obvious from the context they have been translated.

In contrast to the scrappiness and disorder of the collection, the writing itself is usually not too difficult to read. A typical folio is illustrated on the following page. It is p. 106 of Notebook 6 Logic, transcribed as Fragment 7, pages 189-190.

Acknowledgements

For access to the manuscripts, and much help from the library staff, we wish to thank the Royal Society of London. Our information on the Falks in §2 relied heavily on John Rollett, and on the letters to Russell held in the Russell Archives at McMaster University in Canada (archivist Kenneth Blackwell).

For help and advice we thank Desmond MacHale, Daniel D. Merrill, Maria Panteki, Volker Peckhaus and Adrian Rice. Andreas Bächli translated and explained all passages of the Boole material in Greek and Latin.

A special thanks goes to Elaine Lerf. She translated everything written by Gérard Bornet in German: i.e. part 2 of the introduction and much editorial material. She proofread the final copy.

The occupation of Bornet with Boole was part of a broader project on philosophical aspects of the mathematization of logic, which project was supported by the "Swiss National Foundation for the Advancement of

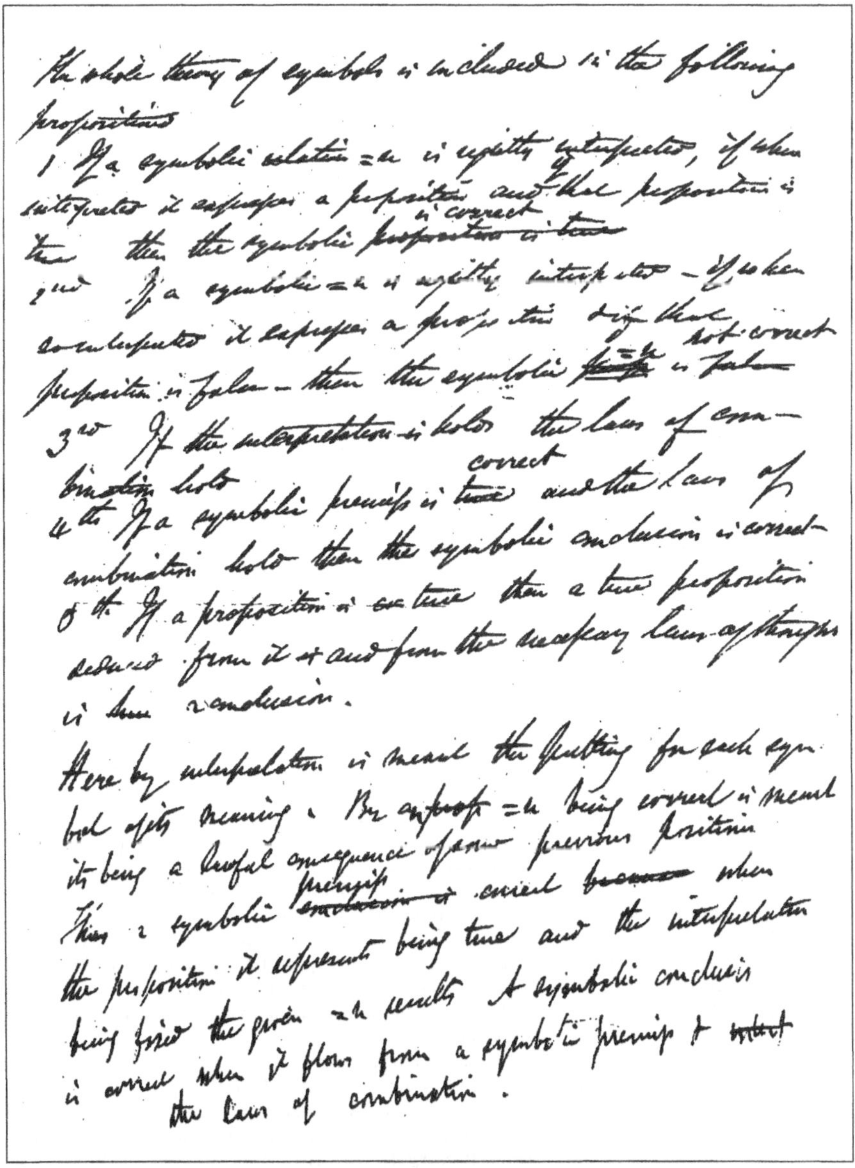

Figure 8: A page from Notebook 6 Logic
(By permission of the President and Council of the Royal Society)
[transcription on pages 189-190]

Scientific Research" to whom thanks are due. We are also grateful for financial support from the "Swiss Society for Logic and the Philosophy of Science", which made a third visit by Bornet to the library of the Royal Society in London possible.

Notes

1 Among biographical writings, the most distinguished is the book-length study *1985a* by the Cork mathematician Desmond MacHale; all further details should be first sought there.

2 Neil's interests included logic and reasoning, which must have drawn him to Boole's work; but it is quite unclear how, being based in Scotland, he came to know so much about Boole, who seems not to have referred to him.

3 In addition, the Stotts seemed to maintain an interest in matters algebraic, including connections to a Liverpool Mathematical Society. Mr. Stott was acknowledged in Ross *1905a*, a re-creation (more or less) of Arbogast's methods which had been discussed before this society.

4 Oswald Toynbee Falk (1879-1972) was a stockbroker who later became involved in Treasury affairs and work with John Maynard Keynes (a younger friend of Russell, incidentally). For context see Skidelsky *1983a*, ch. 14.

5 The full extract reads as follows:

Pour ajourd'hui je ne veux répondre qu'à votre question touchant les mss. de Boole. D'abord, je suis très étonné d'apprendre que ces mss. se trouvent en d'autres mains que celles de Mrs. Boole, qui, m'a-t-on dit, conserve les publication des son mari, et à qui l'on doit s'adresser quand on veut se procurer certaines d'entre elles (c'est ce qui m'est arrivé quand j'ai voulu me procurer quelques mémoires de Boole: mon libraire les a demandés à Mrs. Boole). Quoi qu'il en soit, je suppose que l'authenticité des mss. en question est assurée. Je considérais comme un grand honneur pour moi d'être chargé de l'édition de ces mss., à la condition qu'ils en vaillent la peine (c'est à dire qu'ils ne soient pas de simples brouillons ou copies de ce que Boole a publié), et à la condition qu'ils ne dépassent pas ma compétence (je pense notamment à ce qui regarde le calcul des probabilités, ou je vois pas encore bien clair, et où il est si facile des se tromper: Boole lui-même s'est trompé dans une question capitale de probabilité, selon M. Mac-Coll). Dans ce dernier cas, je demanderais au moins qu'on m'adjoignît un collaborateur scientifique. J'aimerais que ce fût vous, ou M. Whitehead: mais je présume que, si vous me proposez cette tâche, c'est parce que vous ne pouvez pas vous en charger, car vous vous en acquitteriez mieux que personne.

6 Alphonse Gratry (1805-1872), Oratorian Father and graduate of the *École Polytechnique* (a striking combination), published a two-volume *Philosophie — Logique* (1855 and later editions), conveying a mystical and deeply religious attitude to logic, stressing unity, and also covering philosophical questions in mathematics such as infinitesimals and the foundation of the calculus (Gratry *1944a*). According to Mary Boole, George also greatly liked his book, and the features just mentioned would have been congenial for him; but the manuscripts do not confirm her claim.

The contact between Mrs. Boole and Russell may have occurred trough Una Birch (1876-1949), who became a well-known literary figure as Una Pope-Hennessey. She wrote to Russell on 8 December 1904, thanking him for his views on Gratry and contrasting them with the enthusiasm of Mrs. Boole, who hoped for an English translation of *Logique*.

7 On the contributions of De Morgan and Jevons, see §13 and §14 below respectively.

8 In contrast to the great impact of Whately, the treatment of his work and its background in general histories of logic is miserable, and sometimes non-existent. A good start on him can be made with van Evra *1984a* and Dessì *1988a*. On logic teaching in Europe and the USA at that time, see Blakey *1851a*, chs. 14-22.

9 It is possible that Boole and Whately met; in a letter of 15 December 1847 to the Rev. Edmund Larken (an old friend), Boole mentioned that John Graves, Professor of Jurisprudence at University College London and amateur mathematician, was trying to arrange a meeting to discuss *MAL* (University College Cork, Boole Papers, BP/1/223(11)).

10 For some reason Boole did not mention here his paper *1844a* or cite Gregory, who had stated the left-hand trio in the paper *1839a*. In a footnote he allowed (3) to permit "+n = +", which is also to be found in Gregory (*1839b*). In addition, Boole did not attribute the names of the first two laws to F. J. Servois (1767-1847), a follower of Arbogast.

11 On 26 January 1854 Boole wrote to De Morgan, reporting that he had asked his London publisher to send him a copy of *LT*. He also wished to send a copy to Sir John Herschel, and sought his address; and also commiserated with De Morgan on a loss in the family (presumably the death of his eldest daughter Elizabeth Alice the previous month). This letter was given in 1919 by De Morgan's grandson R. Campbell Morgan to the Bodleian Library, Oxford, where it now resides (Ms. Autograph d.14, fols 46-47): it is missing from the edition Smith *1982a* of their correspondence.

12 The identity of Boole's financial friend is not known. Unfortunately no correspondence between the Macmillan Company and Boole is to be found in the publisher's archives in the British Library Additional Manuscripts, 54786-56035, 61894-61896. But the manuscript of *LT* survives in the Royal Society collection (U1-580), except for the last chapter, which is lost.

13 An interesting example is provided by John Hoppus (1789-1879), the founder Professor of Philosophy of Mind and Logic at University College, London. From 1855 until his retirement in 1867 he used *MAL* for a few questions in class examinations but alluded to *LT* only once. He was probably directed to *MAL* by his colleague professor De Morgan, whose own book *Formal Logic* had already been used by Hoppus for questions. His successor, G. Croom Robertson (1842-1892), stopped the practice (information from College Calendars, held in College Archives).

14 Rhees published several bits of this essay in Boole *1952a*, 240-266; the fragmentation has not done it good service, so we give it complete here.

15 A textual puzzle attends this passage from Whately: he wrote 'force' rather than 'form'. Presumably this is a misprint; however, it appears in every edition consulted, including the original encyclopaedia appearance (*1823a*, 209). Boole rendered it as 'form', as quoted here. 'Force' does appear sometimes in writings on logic of this period; for example, De Morgan used it to characterise the intensionality of a term (*1864a*, 105-106, 129).

16 In 1852 Boole wrote a long letter to Hamilton in defence of mathematics and its education; a copy survives in the Cork collection of Boole papers (BP/1/221(1)), and was transcribed in MacHale *1985a*, 133-134. No reply seems to be extant.

17 The word "ideology" comes from "idéologie", proposed in 1796 by Destutt de Tracy to express the triad of the idea, the notion to which it refers, and the sign used to effect the reference. The quite different meaning which it carries today is due to the Marxists.

18 The significance of this passage has been noticed independently in Deakin *1996a*, in a study of some of Boole's methods of solving differential equations. Note also the preface to the book.

19 Boole may have restricted $(x+x)$ to disjoint classes x and y to avoid multi-sets (as they are now called), collections to which a member may belong more than once. The unrestricted form does occur occasionally in Boole, as "signed heaps", to use the phrase of (Hailperin *1986a*). The word "heap" comes from the introduction of multi-sets by the amateur British mathematician A. B. Kempe (1849-1922). His long paper (Kempe *1886a*) was published by the Royal Society, but it interested only C. S. Peirce and, in the 1900s, the American philosopher Josiah Royce (1855-1916); for discussions, see (Vercelloni *1989a*, ch. 1).

20 Venn could only handle up to four predicates with convex shapes; an ingenious algorithmic extension to any finite number is to be found in Edwards *1989a*. On the differences between Euler's and Venn's methods, see Grattan-Guinness *1977a*.

21 Moore *1993a*, xxxvii refers to a letter of 13th June 1902 from Russell, who mentions that after the (putative) completion of his work on *The Principles of Mathematics* (Russell *1903a*) he was undertaking a "systematic study of the great philosophers preparatory to my Logic". In this connection he got to work on *LT* (Russell Archives, ms. 220.010630). The study cannot have been very intensive, for in the space of a few days he also turned to De Morgan *1847a* and Venn *1881a*.

22 Passages where Boole refers to "the actual processes which take place in the mind of a person" (page 56) must be balanced against passages where he makes clear that the intellectual operations he is interested in are not concerned "with the nature of the individual object of thought" (page 67). Accordingly Boole leaves "the material and sensuous element" of concepts out of consideration (page 68). He also refers in his reply to the anticipated objection "that we do not actually reason thus" to an "ideal standard" and remarks, "that the actual performances of our nature" does not "in any case fully answer to its faculties and capacities" (page 111).

23 Here, it is true, Boole states in this connection the modernly assumed criterion that he thinks of "as fundamental those laws and principles from which all other general truths of science may be deduced, and into which they may all be again resolved." (*LT* 5). From today's standpoint the adjective "fundamental" would thereby become relative to an axiom system. The idea of such a relativity was, however, totally foreign at Boole's time, see also §19.

24 Frege (see §14) had Ernst Schröder rather than Boole in mind here (private communication from Volker Peckhaus). Husserl's earlier-quoted statement too was made in connection with a review of Schröder's work. This demonstrates how strongly Boole was considered only as a precursor of Schröder by a later generation.

25 Apart from the concepts of statement and argument, the concept of signs so especially central to formalism cannot be more closely defined without recourse to expressions such as "consciousness" or "thought".

26 Modernly put, Boole in actual fact muses on that which makes up structures without wishing to study structures as such as objects of investigation.

27 This lies behind the question discussed on page 72 as to whether logic is a noetic or ostensive science. The psychologism that is required to be rejected in formal logic sees logic as ostensive science; Boole, however, pronounces in favour of noetic science.

Part A

The Nature of Logic and the Philosophy of Mathematics

Chapter I

The Nature of Logic

————————————[1848]————————————

Logic is the science of reasoning

Reasoning is for the most part carried on by the aid of signs. It has been contended by some writers that it can only be conducted by this agency; others maintain that the use of signs is not indispensable and this is the more probable opinion. But it is universally agreed that the use of signs is a most important aid and that without them no extended process or reasoning could be conducted.

The signs by which we conduct the processes of common reasoning are the words of our own language either spoken or thought. It has been observed that they who are debarred from the use of words are led if capable of reasoning at all to invent a substitute. Laura Bridgman, an American young lady who was born deaf and dumb but was possessed of considerable powers of intellect was accustomed to put her fingers in rapid motion when she was occupied in thought.[1] Dr. Whately has written a tract[2] to prove that it is not the faculty of reasoning but the faculty of reasoning by signs [that] distinguishes man as an intellectual being from the lower animals.

Now logic while it is the science of reasoning in general is in a more especial sense the science of reasoning by signs. It investigates the forms and expressions to which correct reasoning may be reduced and the laws upon which it is founded.

Of what does reasoning consist?

If we examine any regular discourse we shall find that [A 88.1] it consists chiefly of Propositions, that is of assertions either affirming or denying such and such relations of such and such subjects.

When one or more propositions are given and from these we can infer the truth of some other proposition not identical with the given ones such a conclusion is obtained by a process of reasoning, and it is of such processes that we are to give an account.

Now propositions relate to things either real or supposed and by Language we express our conceptions of things. It will therefore be necessary to examine under what aspects things are presented to us, what are the chief varieties to be traced in our conceptions of them and under what conditions we speak of them in common discourse.

General Considerations

Of our Conception of Things

For the first place it may be remarked that we are not confined to the contemplation of things as individuals unconnected with each other but we are able to conceive of them as arranged in classes each of which includes many individuals and under this aspect we designate them by a common or general name. We think and speak not only of Peter and John and other individual men but also of *men* in general and under this conception of *men* we include the particular individuals John, Peter, etc.

In like manner we are able not only to make assertions respecting individuals as John is mortal but also respecting entire classes of individuals as Men are mortal.

In short the world is so constituted as to embrace [A88.2] innumerable individuals possessed of common properties and attributes and the human mind is so constituted as to posses a capacity of forming general ideas and contemplating classes as well as individuals. Finally Language is so constituted as to admit of the expression either of individual things by proper names or of classes of things by common names.

On the relations of individuals and classes

An individual may possess a great variety of attributes and thus belong to a great variety of different classes. The individual John may be considered as belonging to the classes Man, rational being, biped, white, civilized, moral agent etc. The extent to which we are capable of carrying on this classification depends in every instance upon the extent of our knowledge.

To every class there corresponds a contrary class. When we have conceived the class of animate beings we can conceive the entire contrary class of inanimate i.e. not-animate beings, and it is obvious that these two together make up all beings. In like manner the class men and the class not-men together make up all things that exist and similarly do any other class and its contrary class taken together.

It is obvious that a given name and its contrary cannot be applied to any individual. We cannot describe any being as at the same time *man* and *not-man*, *rational* and *not-rational*.

On the other hand either a given name or its contrary may be applied to every individual thing that exists. We may say of anything whatever either that it is a man or that it is something which is not man i.e. either that it is man or not-man. [A88.3]

Of the mode in which the mind combines and modifies its own conceptions

The mind is not a passive recipient of the impressions of external things but it has the power of modifying its own conceptions according to the laws of its own nature.

First, From the conceptions of two distinct classes of things it is able to form the conception of another class as a whole of which those two are parts and it has the power of expressing this aggregation of parts into a whole in Language. Thus from the two distinct conceptions of the class *oxen* and the class *horses* we can form a conception of the aggregate class *oxen and horses* and we can reason upon this aggregate class just as we can reason upon the distinct classes of which it is formed.

Secondly, From the conceptions of two classes of things not entirely distinct we are able to form the conception of a class the members of which shall be common to the two classes which are given. Thus from the two classes of white cattle and horned cattle we are able to form the conception of the class white horned cattle, a class each member of which is at the same time a member of both the original classes.

Thirdly, We have the power of modifying our conceptions by operations inverse to those which have been considered. For as we can form the conception of a whole from the conception of its parts so conversely from the conception of a whole and of one [of] its parts we can form the conception of the other part. Thus if we conceive of men as a whole class and of Europeans as of a partial class included in the former then are we able to conceive of All men except Europeans as the [A88.4] remaining part and we can reason upon this conception just as we can reason upon either of the conceptions originally given. In like manner from the conception of the class white horned cattle already adverted to we can revert by abstraction to the unmixed conceptions of white cattle and of horned cattle. The operations which have here been described both direct and inverse are subject to determinate laws the investigation of which will occupy a future chapter.

The laws of these inverse operations which are derivative and are founded upon the laws of the direct operations already explained and upon a certain other law of direct operation to be thereafter demonstrated will be considered in another part of this treatise.

Of the nature of Propositions

The mind is not only capable of forming conceptions of individual things and of classes of individual things and of modifying those conceptions according to the principles already stated but it is also capable of perceiving relations of equality among things and classes of things the conceptions of which are thus formed and thus modified. The expression of any such relation in Language constitutes a Proposition.

All discourse and reasoning may be resolved into a series of propositions. In Reasoning these flow from each other according to fixed laws which it is the business of the science of Logic to investigate.

The assertion All men are mortal is a Proposition. The assertion Peter is a man is also a proposition. The inference that Peter is therefore mortal constitutes with the propositions from which it is deduced an example of reasoning.

Propositions do not always appear under the form of a [A88.5] relation of equality among things or classes of things but they are always reducible to such a form. The proposition Alexander vanquished Darius is equivalent to Alexander was vanquisher of Darius in which form it expresses a relation of equality between two individuals, that relation being referred to a certain past time.

Of the limitation of the subject of discourse

Both in reasoning and discourse we form a tacit assumption as to the range of things concerning which we reason or discourse.

Ordinarily this assumption amounts only [to] this viz. that the things are things really existing as when we say Men are mortal in which case we speak of all men that exist. Sometimes we implicitly confine ourselves to things existing in a particular country or district as when we say Corn is dear in which case we do not mean all the corn that exists but only the corn that is for sale in a particular region. Sometimes we speak of things which do not exist but are the mere products of the imagination as when we say The Centaur is a fabulous being.

Now whatever is that range of things to which our discourse is confined and from which all the things that we discourse of are taken – that range of things we shall define as the Universe of Discourse.

And this definition being laid down it is clear that the office of a word as *men* is to mark out or select all individuals existing in the Universe of discourse to whom that title may apply.

Suppose for further illustration that we take the following example. The minds of men are agitated during this present year 1848 with aspects [of] political change which timid men view with dread and the ardent with hope. By *men* are not meant *all* [A88.6] *men* – not the savage who never heard of the events referred to – but men who are either witnesses or hearers of them. And when we further limit the individual spoken of by saying timid men, ardent men, by the words *timid, ardent* we mark out of the class before denoted those who possess the particular attributes denoted by the words *timid* and *ardent*.

These instances are confined to the employment of the substantive and the adjective and it is very important to observe that the mental act represented by both is the same in kind. The substantive selects and marks out of an implied universe of discourse the individuals which answer to a given description. The adjective selects and marks out of that class of things denoted by the substantive to which it is prefixed those which answer to the further description which it supplies. In the former case the subject upon which the operation is performed is implied – in the latter case it is expressed.

The articles and other attributives of the singular number represent a similar operation – limiting the subject to an individual with or without further definition.

Furthermore all expressions which serve for the purpose of definition and distinction whether one or many worded represent the same fundamental operation. In the expression The barbarians who subdued Italy the clause "who subdued Italy" is an attributive of this nature. It selects and marks out of the class barbarians those to whom a certain further description may be applied.

Laws of the Mental Operation

The principles which have been established in the preceding chapter may be stated as follows:

1st. The mental operation represented by the adjective [A88.7] the substantive and by all descriptive and attributive expressions is that of selecting from a certain class as subject all the individuals which together answer to a given description.

2nd. If it is the substantive that is employed the subject class is the Universe of discourse and is *implied*, if the adjective is employed the subject class is *expressed* by that word or combination of words to which the adjective is prefixed.

The mental operation here described is subject to certain fundamental and necessary laws which it is next proposed to investigate.

First: Let [us] take the substantive oxen and horses each of which represents the operation of selecting from the Universe of discourse a certain class of beings. It has been seen that by a combination of these results we can form the conception of the compound aggregate oxen *and* horses. In forming the conception of this aggregate class it is indifferent whether we add the conception of the class oxen to that of the class horses or whether we add the conception of the class horses to that of the class oxen. The resulting whole is independent of the order of the component parts. This is a Law of Thought and it is the basis of a corresponding law of Language. Thus as a particular illustration of that law it is indifferent whether we say "horses and oxen" or "oxen and horses". And for the general expression of that law it is obvious that when words relating to different classes of things are connected by the copulative conjunction it is indifferent in what order they are placed. [A 88.8]

If we should represent the class oxen by x and that of horses by y and the conjunction *and* by the sign + (which indeed was originally a contraction of the Latin *et*) we might represent the aggregate class oxen and horses by $x+y$ and the aggregate class horses and oxen by $y+x$. It would then appear that we should have

$$x+y = y+x \qquad\qquad (1)$$

The truth of this equation is however quite independent of the particular interpretation of x and y. If by x and y we represent any other classes of things the same equality will still obtain.

Secondly, From the conceptions of white cattle and horned cattle it has been shewn that we can form the conception of *white horned cattle*.

It is indifferent in this process whether we begin with the conception of horned cattle and limit the things involved under this conception by the further condition of whiteness or whether we begin with the conception of the class white cattle and limit the individual involved under this conception by the condition that they shall be horned. This is a law of Thought and as such it is the basis of a law of Language. The expressions white horned cattle and horned white cattle are equivalent to each other and in general when several terms expressive of equality or in any way serving to limit and define a subject of thought follow each other the order in which they follow is indifferent.

If now we represent the operation of the adjective white by x and the operation of the adjective [horned] by y it is clear that we shall have [A 88.9]

$$xy = yx \qquad\qquad (2)$$

and this relation is true whatever other interpretation of the kind we give to x and y.

Thirdly, It is obvious that the expression white $\overline{\text{oxen and horses}}$ is equivalent to white oxen and white horses. If we represent the operation of the adjective white by x, of the substantive oxen by y and of the substantive horses by z and the conjunction *and* by + we shall have

$$x(y+z) = xy+xz \qquad [3]$$

Again, let x represent the operation of the adjective fruitbearing, y that of the substantive trees and z that of the substantive pear trees then $y-z$ would represent *all trees except pear trees* and by $x(y-z)$ would be represented all of this remainder that bear fruit. Now this result is the same as we should obtain if we should first conceive of all fruitbearing trees xy and then from these take all fruitbearing pear trees xz which would give $xy-xz$ whence we have

$$x(y-z) = xy-xz$$

Fourthly, Let us represent the operation of the adjective good by x then to whatever substantive as men this adjective is applied it marks out and selects the individuals of the class men which answer to the description good. Now suppose the same operation repeated as in the expression good (good men). Here from the class of good men we are supposed to mark out those individuals which are good. It is clear that we mark out the whole and that in fact

$$\text{good good men} = \text{good men}$$

Hence *good* being represented by x we have

$$xx = x$$

and if as in algebra we represent xx by x^2 we have [A88.10]

$$x^2 = x \qquad (4)$$

I call the symbols x, y, z etc. elective symbols because they represent that operation of the adjective the substantive or the attributive clause by which they mark out and select from a given class of things those individuals which possess a given character. This operation is the same in kind in all cases whether the subject class be expressed or only implied.

Comparison of the laws of elective symbols with the laws of arithmetical symbols

[1st.] It has been shewn that if x and y are any two elective symbols then

$$x+y = y+x$$

Now the same equation holds true if x and y represent numbers. We know $7+5 = 5+7$, $6+2 = 2+6$ and generally whatever numerical values we attach to x and y that $x+y = y+x$.

2nd. It has also been seen that x and y being any elective symbols we have

$$xy = yx$$

and this relation is equally true if x and y are any arithmetical symbols. Thus $7 \times 5 = 5 \times 7$, $8 \times 9 = 9 \times 8$ and similarly for any other values of x and y.

3rd. It has also been shewn that x y and z being elective symbols we have always

$$x(y+z) = xy+xz$$

$$x(y-z) = xy-xz$$

This law is also obeyed by symbols of quantity. In fact the expression $x(y+z)$ then denotes the result of multiplying $y+z$ by x which by the rules of algebra is $xy+xz$.

4th. Any elective symbol x satisfies the law whose expression is [A88.11]

$$x^2 = x$$

It is not true that any numerical symbol satisfies this law. Out of the entire range of numerical magnitude from minus infinity to plus infinity there are but two magnitudes which satisfy this law they are 0 and 1. For we have

$$0 \times 0 = 0 \qquad \text{or} \qquad 0^2 = 0$$

and $\qquad\qquad 1 \times 1 = 1 \qquad \text{or} \qquad 1^2 = 1$

These numerical magnitudes may in fact be introduced into the elective system and admit of interpretation there as well as in the system of number to which they also belong.

Consequences of the above analogies

1. It results from the analogies developed in (1) (2) and (3) of the preceding chapter, that we can perform upon elective symbols the same processes in subjection to the same rules as are performed upon the symbols of algebra in the operations of addition, subtraction and multiplication.

2. Since $x^2 = x$ it is plain that $x \times x^2 = xx$ or $x^3 = x^2$, therefore by the previous equation

$$x^3 = x$$

and in like manner $x^4 = x$, $x^5 = x$ etc., whence generally

$$x^n = x$$

From this it follows that if in performing any operation upon elective symbols such a term as x^2, x^3, etc. should appear, it may immediately be replaced by x.

For example if we multiply $x+y$ by $x+y$ the operation performed as in common algebra stands thus [A88.12]

$$x+y$$
$$x+y$$
$$\overline{}$$
$$x^2+xy$$
$$+xy+y^2$$
$$\overline{}$$
$$x^2+2xy+y^2$$

then replacing x^2 by x and y^2 by y we have

$$(x+y)\times(x+y) = x+2xy+y$$

Examples for Practice

Multiply	$x+y$	by	$x-y$	Answer $x-y$
Multiply	$x+2y$	by	$x-3y$	Answer [...]
Multiply	$x+y-xz$	by	$x+z$	Answer [...]
Multiply	$x+y+z$	by	$x-y+z$	Answer [...]

If we multiply x by $1-x$ we get $x-x^2$ which is equal to 0 since $x^2 = x$. In like manner $y(1-y) = 0$ and so on. This result must be borne in mind as it will greatly facilitate the operation of multiplication.

Example

Multiply $x(1-y)+z$ by $y+z(1-x)$

$$x(1-y)+z$$
$$y + z(1-x)$$
$$\overline{}$$
$$yz$$
$$+ z(1-x)$$
$$\overline{}$$
$$yz + z(1-x)$$

Here the term $x(1-y)$ multiplied by y vanishes and the same term multiplied by $z(1-x)$ also vanishes.

Multiply	$x+y(1-z)$	by	$z+1-y$
Multiply	$x-yz$	by	$y(1-x)$
Multiply	$x+2y(1-z)$	by	$z(1-x)+yz$

Of the interpretation of 0 and 1 as elective symbols

It has been shewn that the symbols 0 and 1 satisfy the conditions:

$$0^2 = 0 \qquad 1^2 = 1$$

as well as the other laws of elective symbols. It remains [A88.13] to inquire whether we can interpret them in that system as well as in the system of magnitude.

The classes of things which are presented to our notice vary very widely as to their extent. There are some which consist of many individuals – some of but one or two – and we can conceive of classes which have no existence.

On the other hand a class may be so extensive as to include the whole of existing things and in short to fill up the Universe of Discourse.

The two limits then which bound the possible extent of classes are *nothing* and the *universe*, and there is no difficulty in seeing that the former of these may be represented by the symbol 0 just as we do in arithmetic. We have then only to consider the interpretation of the other symbol 1.

Now it is the characteristic property of 1 in arithmetic that whatever may be the value of any other quantity we have always

$$1x = x$$

and if the symbol 1 is to be introduced into the elective system it must be used in the same manner. The question then is what class must 1 represent in order that the equation

$$1x = x$$

may always be true whatever class is represented by x.

Now by the product of two symbols is represented that class of things which is common to both the classes represented by those symbols. Hence we see that 1 must represent such a class of things that the individuals which are common to that class and to any other class x shall include all that are [A88.14] contained in the class x. Or to state this in other words whatever individuals are found in any class whatever are found also in the class represented by 1.

The class 1 therefore includes all other classes. It is therefore the Universe of Discourse.

Expression of the class not-men etc.

By 1 we represent the Universe of Discourse. Suppose that x represents the class men. Now all the beings in the Universe are either *men* or *not men*. If we take away the class men the remainder will be the class of beings which are not men i.e. the class not-men. Now this taking away is represented by the sign −. Hence if x = the class men then

$$1-x = \text{the class not-men.}$$

And similarly if by y we represent the class plants, by $1-y$ we should represent the class not-plants.

The symbol x by which we have represented the class men satisfies the law

$$x^2 = x$$

In like manner the expression $1-x$ which under the same circumstances represents the class not-men ought to satisfy the condition

$$(1-x)^2 = 1-x$$

and this in fact it does. For if we multiply as below

$$
\begin{array}{l}
1-x \\
1-x \\
\hline
1-x \\
\quad -x+x^2 \\
\hline
1-2x+x^2
\end{array}
$$

and replace in the result x^2 by x we get $1-2x+x$ or $1-x$. Wherefore $(1-x)^2 = 1-x$.

The most general mode of describing any class of things is to particularize certain properties which its members do possess [A88.15] and to particularize certain other properties which they do not possess. We are now in a condition to express classes thus defined. Suppose that by x we represent the class of things that are *white* and by y the class of things that are *square*. Then we shall have the following derived classes.

1st.	xy	Things white and square
[2nd.]	$x\,\overline{1-y}$	Things white and not square
[3rd.]	$(1-x)\,y$	Things not-white but square.
[4th.]	$(1-x)(1-y)$	Things neither white nor square.

To recapitulate what has now been said it appears

1st. That we reason chiefly by the aid of signs.

2nd. That Language as a system of signs is adequate to the expression of our conceptions of things and of the relations which we perceive to exist among the things conceived.

3rd. That things as respects our conceptions of them are presented to us under various forms 1st. As individuals (John) 2nd. As simple classes (men) 3rd. As classes formed by the aggregation of other classes (men and women).

4th. As classes formed by contemplating the individuals which belong to more than one simple class (white horned cattle) and lastly formed by considerations the very opposite of the above as (All men *except* Africans) etc.

4th. That the relations among things are expressed by Propositions and that such propositions may be sufficiently defining the classes of things spoken of [as to] be reduced to expressions of identity or equality.

5th. That when from one or more Propositions another Proposition is deduced as an inference the process of deduction involves an act of Reasoning.

6th. That the same signs are not always used in the same sense but have reference to such individuals answering [A88.16] the description which they convey as are found in the implied Universe of Discourse whether it be 1st. the entire Universe of existing things or 2ndly. a part of that Universe or 3rdly. The range of our own imaginations. [A88.17]

Elementary Treatise on Logic not mathematical including philosophy of mathematical reasoning

—————————————[Probably before 1849]—————————————

The object of Logic as a Science is to explain the laws of those mental operations by which ordinary Reasoning is conducted. The design of Logic as an Art is to exhibit the most useful general forms in which valid argument may be expressed. These objects are perfectly distinct. We might possess an exact knowledge of the ultimate laws of thought without caring to deduce from them the rules of Logic as an Art. On the other hand we might collect by observation a large number of lawful forms of argument, without possessing any acquaintance with the ultimate laws to which as their origin they may be referred.

I design in this treatise to speak of Logic chiefly as an Art, and for the following reasons. First, although Logic has been regarded as a science by different writers it has never assumed a really scientific form. As exhibited in the writings of logicians, it consists of little more than a collection of forms. Secondly Logic as a Science is a branch of the larger science of Reasoning by Signs, another form of which is exhibited in ordinary Mathematics. When the signs which the mind employs relate to things – their attri- [A89.1] butes – and qualities, as do the words of ordinary language, they constitute the basis of Logic. When they relate to the affections of Number, Magnitude and Direction in Space they constitute the basis of Mathematics. Of logic in the scientific form here described, and of the more perfect Art founded upon it, I shall speak in a distinct work. What I here mean to exhibit is so much of the rules and forms of Logic as can conveniently be exhibited without any exact inquiry into first principles, and more especially such of those rules and forms as may be of service in the elementary mathematics.

I design however in doing this to lay down nothing that is inconsistent with the principles of that more general Science of Reasoning of which mention has been made above. I shall without scruple disregard those technical forms and overstep those limitations which have [been] imposed upon the Art of Logic by almost every writer from the days of Aristotle to the present time without at the same time possessing any real foundation in Nature.

Of Signs

In general we reason by the aid of signs. Words are the signs most usually employed for this purpose. It was observed of an American lady who was both deaf and dumb that she aided the processes of thought by means of signs made by the motion of her fingers.[1]

Of Propositions

A proposition is a sentence by which a declaration or [C 26.1] assertion is made, whether absolutely or conditionally as Men are Mortal. If we transgress the rules of Virtue we shall suffer the penalties of Vice.

A proposition is either true or false.

By a primary proposition I mean a sentence which affirms or predicates something of something as when we say Peter is a man. Here it is affirmed of *Peter* that he is *a man*. In this example *Peter* and *a man* [A89.2] are called the terms of the proposition, the connecting word *is* which defines the relation between them is called the *copula*.

Primary propositions may be distinguished according as they either simply affirm the existence or non-existence of the subject or as they assert a relation between two or more classes of things, or between two or more individuals. The simplest kind of primary proposition is that which only affirms the existence or non-existence of the subject as The rhinoceros exists. There is no centaur. Respecting any individual or class of individuals this judgment may be affirmed that that individual or class of individuals either exists or does not exist. This is a necessary proposition.

Of the two terms of a proposition that which is affirmed of the other is called the *predicate*, that term of which the predicate is affirmed is called the *subject*. In the example first given *Peter* is the *subject*, and *a man* the *predicate*.

Both terms of a proposition we shall indifferently call names whether regarded in grammar as substantive or adjective. So any descriptive phrase used as a term of a proposition may be regarded as a *many worded* name as, They who are moderate in their expectation (*subject*) are (*copula*) less liable to disappointment than others (predicate).

Sometimes the copula and the predicate are combined either in single verb or otherwise, as They who sow the wind shall reap the whirlwind. This example may be resolved as follows. They who sow the wind (*subject*) are (copula) they who shall reap the whirlwind (predicate).

Of the Quantity of Terms

In any primary proposition each term expresses either 1st an individual known by a proper name or 2ndly a class of individuals known by some common name or description, or 3rdly a part of such a class whether that [A89.3] part be a single individual of the class, or a number of individuals contained in it.

In the proposition Peter is a man, Peter is a *proper* name applying to an individual, *a Man* is a term applying to some one individual of a class described by the general name *man*.

In the proposition *Men* are *mortal*, the subject Men denotes all the individuals of a certain class of beings to each of [C 26.2] which the name men is applicable. The predicate *mortal* here denotes some out of that entire class of beings to which the term mortal is applicable. We do not by the proposition All men are mortal imply that All men are *all mortals*. There may be, and there are, other mortal beings than men. The complete form of expression would be All men are *some* mortals.

In the proposition Some men are virtuous, the subject Some men denotes a part of the general class Men. The predicate virtuous denotes *some* individuals out of the class of virtuous.

A term is said to be universal when in the proposition in which it occurs, it refers to all the individuals of the class to which either as a descriptive name or as a mere appellative, it applies. If it refers only to a part of the class [A89.4] it is said to be *particular* or *indefinite*. In the Proposition *Peter* is *a man*, the subject Peter is universal, the predicate a man is *particular*. For Peter is the proper name of an individual and it is of that individual i.e. of all that is expressed by the name Peter that the subject speaks. The predicate *a man* is *particular* for it does not refer to all the individuals denoted by the word man, but only to *some* one of them.

In all particular terms the word some is either expressed or understood, and generally in such case it is *expressed* in the subject, and *understood* in the predicate. Thus Some metals are heavy is equivalent to Some metals are (some) heavy bodies the word some being evidently understood in the predicate, while it is expressed in the subject. So the proposition All sheep are animals is equivalent to All sheep are *some* animals.

When by the nature of the proposition a *particular* term is limited to an individual, the word *some* is replaced by the indefinite article *a* or *an* which is equivalent to *some one*. Thus to take our old illustration Peter is a man is equivalent to Peter is *some one man*.

The extent of the meaning of terms considered [A 89.5] as universal or particular is called their quantity. *All men* is a term whose quantity is universal, *some men* a term whose quantity is particular.

When the subject of a proposition is an individual known by a proper name, the subject is said to be singular. It appears from what has gone before that a singular term is universal.

Of the Quality of Terms

The human mind is so constituted that with the idea of any class of things, is suggested the idea of its opposite. The idea of men suggest to us the idea of beings which are not men. To every name and descriptive title a contrary name is conceivable which in some cases already exists in our language and is in all cases easy to be formed by the prefix of the negative particle *not* to the name given. Thus "imperfect" is in common language the contrary name to "perfect", while we might employ the name not-men, as the contrary of men i.e. as the representative of all beings who are not men. [A 89.6]

Two contrary terms will in this treatise be considered as opposed in *quality*, the one being said to be *affirmative* [C 26.3] the other *negative*. Thus *men* being a term of affirmative quality, not-men will be a term of negative quality.

It is a matter of indifference whether we assume a given term or its contrary to have the affirmative character. If men be taken as affirmative, not-men will be negative, and if not-men be taken as affirmative, men will be negative. But for obvious reasons I shall in this treatise consider any term as negative to which the negative particle *not* is prefixed.

It will be observed that the classes expressed by two universal contrary terms are mutually exclusive, i.e. that no individual found in one of them belongs also to the other; secondly that the two classes together make up the Universe. Thus as all beings in the Universe are either men or not men, the two classes *All men* and *All not men* will make up that universe.

Division

Every class of beings admits of division into two subordinate classes to which opposite or contrary names are applicable. Thus the class men may be divided into civilized and not-civilized, men again into white and not-white men. It may at first sight appear that this proposition is not true because we are accustomed in the use of common lan- [A 89.7] guage to the use of terms or names which are not really contrary to each other in the sense here understood. Between "civilized" and "uncivilized" we may interpose the term "half-civilized." But we cannot interpose any term between civilised and not-civilized. For a proposition being either true or false, the term civilized is either applicable to any individual or it is not applicable. But to all examples of the latter case the term not-civilized is by definition applicable.

The division of a class into contraries is called dichotomy and the power of mentally performing this process is of primary importance in the operations of logical inference and is the basis of the most marked distinction between the laws of thought concerned in ordinary reasoning and the laws of thought employed in the analytical operations of mathematics.

Any two or more classes into which a given class is resolved by division, I shall term the *constituents* of that class. Thus if the class Men is divided into Europeans and not-Europeans, then Europeans and not-Europeans will be the respective constituents of the class *men*. Let a second element of distinction be adopted in the term civilized and we [A 89.8] shall have in the whole four constituent classes viz.:

1. Civilized Europeans.
2. Civilized not-Europeans.
3. not-civilized Europeans.
4. not-civilized not-Europeans.

and these constituents together make up the class Men.

Whenever there exist relations among the elements of distinction which are employed in the division of a class, the number of constitutents will be diminished. For example, let us seek the division of triangles with reference to the elements of distinction implied by the words equilateral and isosceles. If those elements were quite unconnected we should have the [C 26.4] fourfold division.

1. Triangles equilateral and isosceles.
2. Triangles equilateral and not-isosceles.
3. Triangles isosceles but not equilateral.
4. Triangles neither isosceles nor equilateral.

But the second of the above constituents has no existence, since all equilateral triangles are isosceles. The first third and fourth would therefore alone remain.

From all which precedes it will appear that the [A89.9] characteristics of a sound division are:

    1st. That the parts should be constituents which secures:
           1st. That they make up the given class.
           2nd. That they are quite distinct from each other, no
           member of one constituent being a member of another
           in other words that the parts do not overlap.
    2nd. That the constituents really exist without violating
           any known or expressed relation among their
           elements of distinction.

These distinctions in the terms of propositions with respect to quantity as universal and particular and with respect to quality as affirmative and negative being premised, we proceed in the next place to consider some among the different varieties of propositions.

Of Some of the Varieties of Propositions

I shall not in this chapter discuss all the forms even of what I have termed primary propositions. To do this would be impossible as their varieties [A89.10] are really unlimited. I intend merely to examine those propositions which arise from the combinations of that variety of terms which has been exhibited in the two previous chapters, i.e. of terms regarded first as universal or particular secondly as affirmative or negative with reference to some one attribute or element of distinction embodied in their name.

And even upon this limited class of propositions the custom of language imposes a certain further restriction. This is that the predicate is always particular. We are allowed to employ the universal sign "All" before the subject of a proposition but it is never either expressed or understood before the predicate. Although it would be a true proposition to assert that All equilateral triangles are *all* equiangular triangles, the conventions of language forbid us to employ such a form, and compel us rather to express the full extent of its meaning by the double system of propositions:

      All equilateral triangles are equiangular.
      All equiangular triangles are equilateral.

propositions which are perfectly distinct and in each of which the predicate is particular. [A89.11]

All the kinds of propositions with which we have to do are illustrated in the following set of examples.

Subject affirmative

1st. All men are mortal.
2nd. All men are not-perfect.
3rd. Some men are civilized.
4th. Some animals are not horses. [C 26.5]

Subject negative

5th. All not-virtuous men are miserable.
6th. All not-true opinions are not-lasting.
7th. Some not-virtuous men are rich.
8th. Some not-virtuous men are not-shunned.

I do not exhibit these as forms to be imitated in ordinary discourse, but as forms to which such propositions in common language as are essentially of the same kind may at once be reduced *whatever elements we assume to possess the affirmative character*. It is usual indeed to confine the attention to the four first forms but that view would be quite insufficient for our purpose.

The conventional use of language assigns to the second and sixth Propositions of the above system [A89.12] a different form. Instead of saying All men are not-perfect we say, No Men are perfect, but the former is the really philosophical form of the proposition. For it is not our intention to speak of a class of beings called "No Men" and to assert of these that they are perfect. On the contrary we really indicate of Men that they are not perfect – that they are *imperfect*. So instead of saying All not-true opinions are not lasting, ordinary language would substitute No opinions which are not true are lasting. But here also the former is the more philosophical expression.*

Although the use of general symbolic forms in an introductory treatise is on the whole to be avoided, yet it may not be undesirable to exhibit in this manner the system of general propositions of which the above is a particular illustration. Its symmetrical character will perhaps be thus better perceived.

Subject affirmative	*Subject negative*
1. All Xs are Ys	1. All not-Xs are Ys.
2. All Xs are not-Ys.	2. All not-Xs are not Ys.
= No Xs are Ys.	
3. Some Xs are Ys.	3. Some not-Xs are Ys.
4. Some Xs are not-Ys.	4. Some not-Xs are not Ys.

---

* In Hebraic constructions this form is often employed, "All they that put their trust in him shall not be confounded".

[A89.13] Here X and Y stand for general names. Thus if we give to X the meaning Men and to Y the meaning mortal being, then the first form above given would represent the proposition:

All men are mortal beings,

which is equivalent to:

All men are mortal.

Propositions whose subjects are universal are called universal Propositions, and Propositions whose subjects are particular are called particular propositions.

In the preceding chapters we have been chiefly occupied with definitions. In the following we proceed to the business of Reasoning or logical deduction. I apply these terms to every instance in which from one or more given propositions another is deduced which differing from them in form is true [C26.6] if they are true. Any one of the given propositions I shall call a *premiss* and any of those deduced a *conclusion*. The process of deduction I shall also term inference. [A89.14]

Of the Limitation of Propositions

By the limitation of a proposition is meant the deducing from it of some proposition less general, as if from the proposition All men are imperfect, we should infer that All good-men are imperfect, or that Some men are imperfect. Such limitation is effected either by dividing the subject or by reducing it from a general to a particular form.

In deducing from the proposition All men are imperfect, the more limited proposition All good men are imperfect we divide the subject. The class Men is resolvable by division into the two classes Good Men, and Men who are not good, and whatever is predicated of the members of the entire class Men is predicated of each member of its constituent division.

In deducing from the proposition All men are imperfect the proposition Some men are imperfect we limit the subject by making it particular instead of universal. This reduction may be referred to the principle already stated. In fact Some men may be regarded as an indefinite constituent part of the class Men, the principle of division being left [A89.15] unstated.

The proposition No men are infallible when expressed in the more philosophical form All men are not-infallible i.e. All men are fallible is by what precedes reducible to

Some men are fallible

by making the subject particular, and to

> All wise men are fallible
> All not-wise men are fallible

by division of the subject.

Thus by dividing the subject of a proposition a series of less general propositions is formed and it may be remarked that the aggregate of such propositions formed on any definite principle of division will be together equal to the proposition from which they are formed. Thus the two propositions

> All wise men are fallible
> All not-wise men are fallible

are equivalent together to the proposition

> All men are fallible.

The general forms of reduction by limitation are these

> 1st. All Xs are Ys. [A89.16]
> Reduced to Some Xs are Ys.

> 2nd. No Xs are Ys (= All Xs are not-Ys).
> Reduced to Some Xs are not-Ys.

Of the Conversion of Propositions

By the conversion of a proposition is meant the putting of the subject in the place of the predicate, and of the predicate in the place of the subject with such change in the quantity and quality of both those terms as will make the resulting proposition a logical consequence of the given one. Such a conversion is in the language of logicians said to be *illative*, i.e. obtained by *inference*.

There are really but two forms of *illative conversion*, one applying to particular, the other to universal propositions.

If we take such a proposition as Some English men are poets, it is obvious that we may from this at once infer that Some poets are Englishmen. This [A89.17] is the simplest kind of illative conversion. In the example we have chosen, both the subject and the predicate are *particular and affirmative*: Some Englishmen, (some) poets; and in the converted proposition they are *particular* and *affirmative* also – *some poets*, (some) Englishmen. [C26.7] The only difference is that the *some* which is attached to Englishmen in the one case, is attached to poets in the other in accordance with that custom of language already adverted to which confines the *expression* of the word Some to the subject of a proposition.

From the proposition "Some Englishmen are not brave men", we may infer that "Some who are not brave men are Englishmen". This is another example of the illative conversion of a particular proposition and it is here also seen that the quantity and quality of both terms remain unchanged.

From such examples we may establish the following Rule.

Particular propositions may be converted without changing the quantity or quality of either term.

Let it be observed that from the particular proposition Some Englishmen are not brave, we cannot infer [A89.18] that Some brave men are not English, although the proposition is unquestionably true. The student must be careful to distinguish between propositions which are true in themselves and conclusions which are *correct* by way of *inference* from their premises.

From the proposition that All men are mortal we may infer that All who are not mortal are not men. This is an example of illative conversion. In the original proposition the subject is universal affirmative, the predicate particular affirmative, but in the result obtained by conversion the quantity and quality of both the original terms are changed. Instead of (some) mortals we have All not-mortals, and instead of All men we have (some) not-men, the conversion in fact being

From,   All men are mortal.
To,   All not-mortal are not-men.

Thus the original predicate with quantity and quality both changed, forms the new subject, and the original subject with quantity and quality both changed forms the new predicate. [A89.19]

Let us next take the universal proposition No men are perfect beings. We may obviously infer from this that No perfect beings are men. Now the real meaning of our original proposition is that

All men are not-perfect

and the real meaning of our conclusion is that

All perfect beings are not-men.

Here it is also seen that the quantity and quality of both the original terms have been changed in the process of conversion.

From all which we may establish the following Rule.

A universal proposition is converted by changing the quantity and quality of both the terms which it involves.

It is to be observed that without such change of the quantity and quality of both terms, a universal proposition cannot logically be converted. From the proposition All men are mortal, we cannot infer that All mortals are men. So neither from the proposition, All equilateral triangles are equiangular can we infer that All equiangular triangles are equilateral, although the latter proposition expresses a familiar truth. [A89.20]

Formal examples of the above rules are exhibited in the following scheme. [C 26.8]

Propositions	*Conversions*
1st. Some Xs are Ys	Some Ys are Xs.
2nd. Some Xs are not-Ys	Some not-Ys are Xs
3rd. All Xs are Ys.	All not-Ys are not-Xs
4th. No Xs are Ys	No Ys are Xs.

Note. In a proposition in which both terms are singular, and therefore both universal, conversion takes place according to both rules above stated. Thus from the proposition

Victoria is our queen

we infer both the conclusions.

Our queen is Victoria
Whoever is not our queen is not Victoria.

It may also be remarked that in many cases a proposition is transformed both by limitation and conversion. Both these processes are employed in deducing from the universal proposition All ores are minerals the particular converse Some minerals are ores.

Of Syllogism

The nature of a syllogism will be best explained by a few familiar examples.

If we combine the premises [A 89.21]

All men are mortal
Peter is a man

we may from these at once draw the conclusion

Peter is mortal

and the premises and conclusion together constitute a syllogism.

In the above example, we notice that there are two terms in the premises namely, Peter and mortal which reappear in the conclusion, and two terms viz. *All men* and *a man* which do not make their appearance in the conclusion. Let us agree to call the two former terms the extremes, and the two latter terms the middle terms. It is seen that the extremes involve different elements, the middle terms the same element.

A syllogism may then be characterized as follows

1st. It consists of three propositions viz.: two premises and a conclusion drawn from them.

2nd. Each premiss involves two terms viz.: An extreme and a middle term, the middle terms being such as to involve some common element and then connect the premises together.

3rd. The conclusion connects the two extremes so changed if necessary in their quantity and quality as to make it a logical consequence of the premises.

Let us take as a second example the regular syllogism. [A 89.22]

Premises   All fishes inhabit the water
            Some animals which have scales do
            not inhabit the water

Conclusion Some animals which have scales are
            not fishes.

In any such case as the above it is desirable first to resolve the verbal part of each proposition into a copula and an ordinary term. Thus we may write our syllogism as follows:

Premises   All fishes are inhabitants of the
            water.
            Some scaly animals are not inhabit-
            ants of the water.

Conclusion Some scaly animals are not
            fishes. [C 26.9]

The middle terms are "inhabitants of the water" and "not inhabitants of the water", both of which are particular, but while the former is affirmative in quality, the latter is negative.

The extremes in the premises are "some scaly animals" (particular affirmative) and "All fishes" (universal affirmative). The corresponding extremes in the conclusion are "Some scaly animals" (particular affirmative) some not fishes (particular negative). Here then while one extreme remains unchanged in the conclusion, the other is changed both in quan- [A 89.23] tity and quality. In the syllogism first discussed both extremes were unchanged in the conclusion.

The object then of a practical analysis of the syllogism is to be determined.

1st. Under what conditions syllogistic inference is possible.

2ndly. How the quantity and the quality of the extremes will be affected in the conclusion.

I say *syllogistic* inference in the above case because from any pair of premises and indeed from any single premiss other inference is possible. [C 26.10]

Conditions and Rules of Syllogistic Inference

It has been seen that in the premises of every syllogism there occur two middle terms, which serve as connecting links between them. Now the nature of the connexion which they establish will depend upon the question whether those middle terms are of like quality or of unlike quality.

This will be evident from the following consideration. If the middle terms are of like quality [as] All men, Some men, they relate to individuals of the same class, and it is by the connexion which [A89.24] the extremes severally bear to this common class that their mutual connection is determined. But if the extremes are of unlike quality, they relate to individuals of contrary classes so that there is no direct medium of comparison between the extremes. In such cases it is only by conversion of one of the premises that the required medium of comparison can be established. For it has been seen that in the conversion of a universal proposition, the quantity and quality of both its terms are changed. Hence if the middle terms are unlike, and one of the premises is universal, we can by converting that premiss reduce the middle terms to likeness of quality.

In what follows I shall speak of classes or parts of classes as *equal* when the members of the one are members of the other. In this sense of the word equal, as well as in its mathematical sense it is an evident axiom that Things which are equal to the same thing are equal to each other.

Now if we take the syllogism

All men are mortal
Peter is a man
Therefore Peter is mortal.

we see that in the first premiss an equality is [A89.25] asserted between the entire class All men and some members of the class *mortal*. In the second premiss an equality is asserted between the individual Peter and an individual member included in the former class All men. It follows hence that there exists an equality between the individual Peter and a member of the class mortal. In other words the two extremes being compared, the one with an entire class All men, the other with a part of that class, it follows, that they may be compared with each other. The rule of inference in this case is that we equate the extremes i.e. assert an equality in the sense above explained between them.* [C26.11]

---

* The principle which that rule embodies is that whatever is asserted of a whole class is asserted of any collection of individuals contained in that class.[2]

And the rule applies whenever the middle terms are of like quality and one at least of them universal.

Let us in the next place suppose that the middle terms are of unlike quality. Two conditions of inference then present themselves.

The first is when either extreme is universal. This condition is exemplified in one of our previous syllogisms the premises of which are

All fishes are inhabitants of the water.
Some scaly animals are not inhabitants of water,

in which the extreme, All fishes, is universal. [A89.26]

If we convert the premiss in which that term occurs we have the following pair of premises.

All animals which are not inhabitants of the water are not fishes.
Some scaly animals are not inhabitants of the water.

The two middle terms now are All not-inhabitants of the water, and (Some) not inhabitants of the water, and as they are alike in quality, while the former of them is universal, we have, on equating the extremes by the previous rule:

Some scaly animals are not fishes.

In this way any example of the above kind may be treated. But the conclusion will more readily be obtained by simply changing the quantity and quality of one universal extreme and equating the result with the other extreme.

The second condition of inference is when both middle terms are universal. The following premises will serve as an example.

All who are good are deserving of solid esteem.
All who are not good are not truly wise.

Here the middle terms are unlike, and both are [A89.27] universal.
By the conversion of the first proposition our premises become

All who are not deserving of solid esteem are not good
All who are not good are not truly wise,

the middle terms of which are alike, and one of them is universal, whence by equating extremes we have

All who are not deserving of esteem are not truly wise,

as our conclusion.

Suppose in the above case that instead of converting the first premiss, we convert the second. The given propositions then are,

All who are good are deserving of solid esteem.
All who are truly wise are good.

Here the middle terms are alike, and the former of them is universal. Hence equating extremes we get the conclusion:

All who are truly wise are deserving of solid esteem.

Now this conclusion is equivalent to the one before obtained, for the one is in fact the conversion of the other.

Thus also may any similar pair of premises be treated. But the conclusion will be more readily obtained by simply changing the quantity and quality of either extreme, and equating the result with the other extreme. We will now recapitulate the rules [A89.28] at which we have arrived. [C26.12]

1st. Case: The Middle terms of like quality.
Condition of Inference. One middle term universal.
Rule. Equate the extremes.

2nd. Case: The middle terms of unlike quality.
1st. Condition of Inference, one universal extreme.
Rule. Change the quantity and quality of that extreme and equate the result with the other extreme.
2nd. Condition of Inference: Both middle terms universal.
Rule. Change the quantity and quality of either extreme, and equate the result with the other extreme.

These are the ultimate and sufficient rules of syllogism, and they are the only exact system of such rules that has been published.

It will be useful to apply these rules to some general forms.

Example 1st.   All Ys are Xs
                 All Zs are Ys

This belongs to Case 1st. All Ys is the universal middle term. The extremes equated give

All Zs are Xs.

as the conclusion. Note that the only restriction upon the order of the terms in the conclusion is that a universal term if there be one must be made the subject [A89.29] since the predicate is particular.

Example 2nd.   All Xs are Ys.
                 No Zs are Ys.

These premises are equivalent to

All Xs are Ys
All Zs are not-Ys

They belong to Case 2, and satisfy the first condition. The middle term is particular affirmative in the first proposition, particular negative in the second. Taking All Zs as the universal extreme, we have on changing its quantity and quality Some not-Zs and this equated to the other extreme gives

All Xs are (some) not-Zs
Or No Xs are Zs

as our conclusion. If we take All Xs as the universal extreme we get:

No Zs are Xs

as our conclusion. These results are equivalent by conversion.

Example 3rd.   All Ys are Xs.
All not-Ys are Zs

This belongs to Case 2 and satisfies the second condition. The extreme Some Xs becomes on changing its quantity and quality All not-Xs whence we have the conclusion.

All not-Xs are Zs [A89.30]

The other extreme similarly treated would lead us to the conclusion

All not-Zs are Xs

which is equivalent to the former one by conversion.

Example 4th.   Some Ys are not-Xs
No Zs are Ys.

which is equivalent to

Some Ys are not-Xs
All Zs are not-Ys.

This belongs to Case 2 and satisfies the first condition. The result is

Some not-Zs are not Xs [C 26.13]

or its equivalent by conversion

Some not-Xs are not-Zs.

There are two forms of argument depending upon the syllogism of which it remains to give an account. They are the enthymeme and the sorites.

The enthymeme is a syllogism of which one premiss is not expressed, but understood or taken for granted as, Corn is scarce, Therefore its price is high, the suppressed premiss being that commodities which are scarce secure a high price.

A sorites is a series of premises in which the predicate of one is made the subject of the next followed by a conclusion which connects the subject of the first with the predicate of the last as,

The possession of genius confers power.

They who possess power have the means of benefiting their fellow creatures.

They who have the means of benefiting their fellow creatures are bound to exercise those means.

Therefore: They who possess genius are bound to exercise their means for the benefit of their fellow creatures. [A 89.31]

General Remarks on Primary Propositions

In Propositions which express relations among classes, we not only meet with the varieties above described, but an infinite variety of others in the expression of which the conjunctions *and, either, or* etc., are employed. The water consists of lakes, rivers and seas. Plants are either trees or herbs etc., are examples of this kind. For the general treatment of such propositions it is impossible to lay down any system of rules at all analogous to those which have been exhibited above. They nevertheless constitute the objects of a true science – a science which is mathematical in its forms and expressions and which, setting out from a few elementary laws of thought, establishes General Methods adequate to all the purposes of analysis and of inference. The doctrines which have been established in the preceding pages do not constitute any part of the foundations of that science but they are consistent with it and were in fact for the most part derived from an interpretation of its results.

Of Secondary Propositions

By secondary or as they might be termed reflex propositions, I mean propositions which express a judgment with respect to other propositions, and this they may do in two ways. Either they express a judgment upon some single proposition by asserting its truth or its falsehood, or they express a judgment respecting two or more propositions by asserting the existence of some relation or dependence among them.

When we say It is true that virtue promotes happiness, It is not true that utility is the ground of moral obligation we state propositions of the former kind. When we say, If virtue promotes happiness, then self-interest confirms the decisions of conscience, we state a proposition of the latter kind. We express a relation between [A 89.32] the primary propositions "Virtue promotes happiness" and "Self-interest confirms the decisions of conscience." [C 26.14]

The distinction here noticed is of the same kind as that which was adverted to in ( ). And indeed both the varieties and rules of secondary propositions have an exact correspondence with those of primary propositions. The difference which exists is in the subjects of thought, not in its formal laws. I do not however intend to pursue the analogy of its minuter details here.

From what has preceded we may remark that every primary proposition gives rise to two secondary propositions, the one asserting its truth, the other asserting the falsehood of a proposition contrary to the given one. Thus the proposition All men are mortal gives rise to the two reflex propositions It is true that All men are mortal. It is false that Some men are not mortal. To this branch of our subject we may therefore refer the necessary relations which exist among primary propositions. By necessary relations I here understood those relations which are necessitated by the laws of Thought.

In this sense the propositions All men are mortal. Some men are not mortal are said to be contrary to each other. If we compare them we see [A89.33] that their subjects differ in quantity and their predicates in quality. And thus to all the propositions of the systems given in ( ), the following rule may be applied.

From a given proposition to deduce its contradictory:

Rule. Change the quantity of the subject and the quality of the predicate.

Thus the contradictory of the proposition No men are perfect, would be Some men are perfect, and the contradictory of the proposition Some men are not selfish would be All men are selfish.

If we are ignorant respecting a given primary proposition whether it is true or false, we can yet form of it this judgment, viz. that it is either true or false. This is a necessary truth, and it is analogous to that necessary truth established in ( ) that a thing either exists or does not exist.

Secondary propositions which relate to more than one primary one are usually expressed by aid of the particle *if* or the particles *either* and *or* but they in all cases admit of reduction to another form which will be noted hereafter.

Propositions which are expressed by the particle [A89.34] if are denominated conditional propositions.

Ex. If two straight lines intersect, the angles which are vertical to each other will be equal. Here the proposition Two straight lines intersect is called the antecedent, and the proposition The angles which are vertical to each other will be equal is called the consequent.

A hypothetical proposition of the above kind is similar to a universal primary proposition, and is converted in a similar manner. Ex. If the clouds shall have broken, the sun will shine, is converted into If the sun shall not shine, the clouds will not have broken.

In general let X and Y represent the antecedent and the consequent. Then a given hypothetical proposition is of the form

If X is true Y is true

and its conversion is of the form

If Y is not true X is not true. [C26.15]

The proposition Sometimes if X is true, Y is true might be termed a particular hypothetical proposition, and like the particular [A89.35] propositions already discussed, it may be converted without negativing either of the terms, the converse being Sometimes if Y is true X is true.

Various forms of argument to which particular names have been given are subjoined.

Constructive Hypothetical Syllogism

If X is true, Y is true. But X is true, Therefore Y is true.

Destructive Syllogism

If X is true, Y is true, But Y is false, Therefore X is false.

Disjunctive Syllogism

Either X is true or Y is true, but X is not true, therefore Y is true. [A89.36 / C26.16]

Simple constructive Dilemma

If X is true Y is true and if Z is true Y is true
But either X is true or Z is true
Therefore Y is true.

This however is an exceedingly imperfect account even of those forms of reasoning which it professes to examine. It has been observed that ultimate laws of secondary propositions are in perfect analogy with those of primary propositions. So likewise are the forms of Inference. To shew this in syllogism: Let the antecedent and consequent X and Y be called the terms of the proposition and let them be similarly affected with quantity and quality. The four following forms will then be analogous to the four first forms of ( )^3

If X is true Y is true
If X is true Y is not true
Sometimes if X is true Y is true
Sometimes if X is true Y is not true

the terms being respectively of the same quantity and quality in each of the above as in the proposition with which it is compared. The forms of hypothetical will be co-extensive with those of ordinary syllogism and the rules will be the same in substance, with this difference only in expression that instead of saying "equate the extremes" we must say write them as

[A90.1] antecedent and consequent, the universal term when there is one becoming the antecedent and the word "sometimes" being expressed when the antecedent is particular. Ex[ample]

> If X is true Y is true
> Sometimes if Z is true Y is not true

Here the middle terms are unlike and one extreme is universal. Condition 2nd. Case 1. Changing the quantity and quality of that extreme and connecting it with the other extreme, we have indifferently

> Sometimes if Z is true, X is not true
> Sometimes if X is not true Z is true

It is needless to multiply examples.

Conditional propositions can always be reduced to the disjunctive form and conversely disjunctive propositions can always be reduced to the conditional form.

From the conditional proposition If the sky is clear the sun shines we may deduce the disjunctive proposition Either the sky is clear and the sun shines, or the sky is not clear, or we may deduce the disjunctive proposition Either the sky is clear, or the sky is not clear and the sun does not shine. And from either of these (equivalent) disjunctive propositions we may infer the conditional proposition from which they were derived.

Every secondary proposition whether disjunctive conditional, or of any other form may be resolved into a set of denials or negations. Thus from any of the three propositions which have been given above we may deduce the negation It is not true that the sky is clear and the sun does not shine. [A90.2]

These remarks are useful not only as shewing the relations which exist among the different kinds of propositions, but also as indicating that our power of conceiving of these kinds is not to be referred to distinctive faculties of the mind or to distinct conditions of thought as it seems to be thought by some modern writers.

Of the Demonstrations of Geometry

The great divisions of modern mathematical science are Geometry and Analysis the latter being distinguished from the former not so much by the subjects of which it treats as by the use which it makes of symbols.

The demonstrations of Geometry are a particular application of Logical argument, to questions of space, magnitude and figure. As all reasoning consists in the deduction of conclusions from premises so geometrical reasoning consists in deducing geometrical relations from geometrical premises which premises are either truths established by previous reasoning or they are the axioms and definitions upon which all geometry rests.

A definition explains what the thing defined is – and so distinguishes it from all other things. Hence a definition must be convertible with the name of the thing defined. Thus the definition of an isosceles triangle as a triangle having two equal sides, serves to distinguish the isosceles triangle from all other geometrical figures. And as that definition is applicable to no other thing it follows that a triangle having two equal sides is an isosceles triangle. The definition "a triangle having two equal sides" is therefore [A90.3] convertible with the name "isosceles triangle".

In this way Geometry defines the different varieties of lines angles, surfaces etc., with their mutual relation to each other as parallel, perpendicular etc. Such definitions are properly nominal because they are in reality definitions of the meaning of terms.

Axioms (ἀξιώματα)[4] are self-evident truths. We cannot fix our attention upon the conceptions of geometry without perceiving that certain relations are universally true independently of any process of argument. Such truths are called first truths necessary truths or axioms – first truths because they serve as the starting points of demonstration, necessary truths because the mind recognizes them not only as true but as necessarily true, – *axioms* because they are *worthy* of acceptance *per se*. That two straight lines cannot intersect in more than one point is a truth of this nature. It requires no demonstration but on the other hand serves as a basis upon which other truths may be established.

The axiom that Things which are equal to the same thing are equal to each other is an axiom of this kind. The mind cannot avoid perceiving its truth but it does not derive that perception from contemplating a definition.

It has been much disputed whether our knowledge of the axioms of geometry is founded upon experience alone i.e. from observing that they are true in every particular instance, or whether the observation of a single instance awakes the perception of a truth already existing in the mind. This question does not belong however either to Logic properly so called or to geometry. [A90.4]

Note:[5] Perhaps it may be true that all axioms are dependent upon laws of thought and ought rather to be referred to the organic constitution of the mind itself than to its outward conceptions. The proposition $a+b = b+a$ in arithmetic is a truth which has that [?] nature and so are the fundamental laws of words as symbols upon which Logic as a science depends.

There is nothing unreasonable in the supposition that the mind as an instrument may be subject to laws by which the character of its operations is determined. Whatever truths then are immediate consequences of those Laws will assume a character of *necessity*. This view is equally remote on the one hand from the doctrine that there are relations among geometrical figures independent of the existence of a perceiving and conscious mind and on the other hand from the doctrine that the mind is furnished with a store of necessary truths to the consciousness of which it only awakes as experience furnishes occasion.

To exhibit this view more fully let us examine one or two distinct mental operations and endeavour to ascertain the laws to which they are subject.

The adjective in ordinary language may be regarded as the sign of a mental operation Let us consider the term. By men we understand all the members of a certain class possessing given characteristics: animal life, rationality, the male sex etc., by good men we understand a certain portion of that class the members of which possess another characteristic defined by the word "good". The adjective then in its actual use represents the operation of selecting from a class all the members which possess a given character and this operation is of such a nature that if it be [A90.5] repeated any number of times, the effect is the same as if it be performed only once. The term good good men is equivalent in meaning to good men. This may be considered as a law of the operation in question.

Again we can by a certain mental operation pass from the idea of an effect to the idea of a cause of that effect and this operation is subject to a law of thought of which we find an expression in the formula:

Cause of Cause = Cause.

If A is the cause of B and B a cause of C then A is a cause of C. The above formula may be regarded as expressing a law to which the mind is subjected whenever it deals with the conceptions of causality.

Now to pass to the geometrical axiom more immediately under consideration it may reasonably be conceived than whenever the mind has to do with the conception of equality its operations may be subject to a law the actual character and expression of which will be suggested to us by experience. And such a law may I think be detected in the axiom: If A is equal to B and B equal to C then A is equal to C. Perhaps the truth – If a line A is parallel to B and B parallel to C then A is parallel to C may equally [be] referred to a law of the mental conceptions.

It is characteristic of all truths of the above kind that they are made as evident to us by a single instance as by a thousand and that when once seen they are seen to be necessary and universal. It is not so with the laws of the material Universe our knowledge of which rests upon innumerable observations and as such can never be emancipated from the [A90.6] conditions of probability – high as the measure of that probability may be. May we not assume it as a definition of necessary truth that it is such truth as has its origin in the laws of the mind itself? Is it not clearly in this sense that the conclusions of deductive reasoning are said to be *necessary* consequences of the premises? Necessary because the mind cannot abrogate the laws of its own constitution – cannot in fact change the conditions of its own activity. [*End of note.*][5]

The successive steps in the demonstrations of geometry are usually syllogistic e.g. Euclid Prop. 1.[6]

> All the radii of the same circle are equal.
> But AB and AC are radii of the same circle.
> Therefore AB and AC are equal.

and again in the same proposition.

> Things which are equal to the same thing are equal to each other.
> But AB and CB are equal to AC.
> Therefore AB and CB are equal to each other.

The general form of such arguments is

> All Ys are Xs
> All Zs are Ys
> All Zs are Xs.

All the conclusions of geometry exhibited in the form of propositions are universal i.e. they have universal subjects.

Ex. All triangles on the same base and between the same parallels are equal. All right angled triangles have this property that the square of the hypothenuse is equal to the squares of the base and perpendicular etc. The general logical type of such proposition is

> All Zs are Xs

of which the logical conversion is [A90.7]

> All not-Xs are not-Zs.

But it very commonly happens in geometry that the proposition All Xs are Zs which is called the converse of the proposition All Zs are Xs is also true.

Thus the converse of Prop. 5 Bk. 1. The angles at the base of an isosceles triangle are equal is established in Prop. 6 viz.: If the angles at the base of a triangle are equal the opposite sides are equal. It does not however always happen that the converse of a geometrical truth is equally true. It is true that Similar polygons have their corresponding sides proportional. It is not necessarily true that if the corresponding sides of two polygons are proportional those polygons are similar.

The converse of a given proposition in geometry when true is usually proved by an indirect demonstration called *reductio ad absurdum*. This consists in assuming that the thing to be proved is not true and deducing from that assumption some conclusion which is manifestly false. Whence it follows that the thing to be proved is true.

This method of demonstration depends on the principle that a false conclusion implies an error in the premises – a principle which is in fact exemplified in the logical conversion of the proposition:

If X is true Y is true

into the proposition

If Y is not true X is not true.

in which X represents the supposed false premiss or assumption, Y the supposed conclusion drawn from it.

Euclid sometimes employs the disjunctive form of argument [A90.8] e.g. Prop. XIX Bk. 1. The line AC is either less equal to or greater than AB. But it is not less than AB and it is not equal to AB. Therefore it is greater than AB.

And he often proves a general theorem by proving it in all the distinct cases which are conceivable. Ex. Prop. XX Bk. 3. This form of inference is called a perfect induction or an induction per enumerationem simplicem. But it would be better to regard it as a species of deduction. The word induction is in the present day applied to another and higher process of thought – the discovery of universal truths from a limited number of examples by detecting the general law of their connection.

Of the Foundations of Algebra

Algebra in its ordinary acceptation may be defined as a system of reasoning carried on by signs and characterized by the following marks.

1st. The signs are of two kinds, viz.: [Firstly] those which represent numbers and Secondly, those which represent the operations which we are capable of performing upon numbers.

2ndly. The symbols which represent numbers are either particular as 2, 5, 20, etc., or general as a, b, x, etc. and in both cases they are subject to laws of combination founded upon the laws of combination of the things signified.

3rdly. The signs by which operations are represented are [A90.9] subject to laws upon the nature of those operations and these laws constitute the axioms of the science.

Let us endeavour to illustrate these positions. Algebra then in its ordinary acceptation is a science which relates to Number. If the numbers are all known and particular and it is only required to perform upon them certain definite operations according to a rule the science merges into Arithmetic which is in fact a particular Algebra or rather a particular Art founded upon the Science of Algebra. On the other hand if the symbols of number are general and we either seek to perform operations upon these or from given relations existing among them to deduce others even as in logic we deduce conclusions from premises the science we employ is that of

Algebra in the larger sense applied to it by Newton of a Universal Arithmetic.[7]

In general, Algebra investigates relations and prescribes general rules. Arithmetic makes particular applications of those rules.

We see in Arithmetic that $7 \times 8 = 8 \times 7$, $5 \times 3 = 3 \times 5$ etc. A single example of this kind suggests to us the general principle that if two numbers are multiplied together the order of multiplication is indifferent. Algebra expresses this truth in the general theorem:

$$ab = ba$$

which exhibits in fact a law of combination of the symbols a and b equally applicable whether a and b have integral or fractional values.

Other laws of combination for algebraic symbols are expressed in the forms [A90.10]

$$a + b = b + a$$
$$a(b + c) = ab + ac$$

etc. And such elementary laws of the symbols are the foundations of the rules of algebraic addition multiplication etc.

Similar in origin to these are the laws of signs

$$+a \times -b = -ab$$
$$-a \times -b = ab$$

Beside the above Algebra recognizes certain general axioms which are common to it and Geometry such as If equals be added to equals the results are equal. If equals be taken from equals the remainders are equal etc. Perhaps these axioms may be generalized into the following "If equivalent operations be performed on equal subjects the results are equal" and may be considered as expressing a law of thought to which the mind is subject when dealing with the relation of equality.

The operations of Addition and Subtraction may be considered as inverse to each other – so likewise may the operations of Multiplication and Division.

A given subject produces a certain result; the inverse operation performed upon the result reproduces either the subject as in Algebra or a more general result comprising that subject – as in some other branches of mathematics.

The very idea of an operation suggests the idea of an operation inverse to itself i.e. of an operation which has the power of undoing what the other has done. A direct operation performed upon a given subject produces a given result, the inverse operation performed upon that result reproduces either the subject given as in Algebra or a more general result comprising that subject as in the Integral Calculus. The operations of addition and subtraction those also of multiplication [A90.11] and division may thus be considered as mutually inverse.

Thus the elements which are expressed by the symbols of algebra are

1st. Numbers whether integral or fractional, whether particular or general.

2nd. The direct operations of Addition and Multiplication and the corresponding inverse ones.

3rd. The relation of equality.

And upon these simple elements variously combined the entire superstructure of the science[8] is based.

The rules and methods of algebra once established, it becomes applicable to all sciences in which either number or anything which is capable of being represented by number is concerned. Thus in Geometry a given line being assumed as the unit of linear measure, suppose 1 foot, any other lines may be referred to it as a standard of length and the length of that line may be represented by the number of times which it contains the length of the given line. And a given surface as a square foot being assumed as the unit of square measure the area of any other surface may in like manner be represented by the number of times which it contains the given surface. And similarly for solids. Now the elementary propositions of Geometry express the relations among the lengths of lines or between the lengths of lines and the areas of surfaces. Hence they imply relations between the numbers by which the lengths of those lines or the areas of those surfaces are expressed, a given length or a given area being the unit. Those relations may be expressed in the language of algebra and new truths deduced from them by the methods of algebra. This gives rise to the Cartesian method of the "Application of Algebra to Geometry". [A90.12]

In like manner in the Science of Mechanics a given weight being assumed as a unit any other weight or pressure may be represented by the number of units of weight which would be equivalent to it. So also may Force which is the cause of Weight, and Velocity and Time and other elements be expressed by symbols of Number.

And thus as the forces and energies of the material Universe, the effects which they produce, and the conditions of Time and Space under which they operate have been made subject to numerical evaluation, the sciences of Algebra and Geometry are indispensable prerequisites to the study of its phenomena.

Of the Differential Calculus

If the preceding chapter has been understood there will be little difficulty in comprehending the fundamental principles of the Diff[erential] and Int[egral] Calculus which are in fact an extension of ordinary Algebra, that extension depending upon the introduction of a new Idea – the Idea of a Limit.

In ordinary algebra we have to do with fixed quantities alone, i.e. with quantities whose values are fixed and constant whether we know what those values are or not. In the differential calculus we regard them as varying and in that variation passing from one magnitude to another through all intermediate stages of magnitude. Now this is in accordance with observation. A stone falling by the force of gravity has at the instant of its reaching the earth a certain velocity. But in order to reach that velocity it has passed through all inferior stages of velocity.* [A90.13]

While the stone is falling the space which it has described continues to increase. But it does not increase uniformly with the time. The rate at which it is increasing at any instant is measured by its velocity at that instant, a velocity which was less in any preceding instant *however near* and will be greater in any subsequent instant *however near*.

Now the conception of this varying velocity or varying rate of increase involves the idea of a Limit.

If the velocity were uniform we could express it numerically by comparing any space described with the time in which it was described. Let 1 foot be the unit of space 1 second that of time and let a uniform velocity of 1 foot per second be called the unit of velocity and be represented by 1 then a velocity of three feet per second would be represented by 3 a velocity of three feet in two seconds by $\frac{3}{2}$ and in general any velocity would be found by dividing the number of feet described by the number of seconds in which it was described. [A90.14]

But if the velocity were constantly increasing as in the case supposed and we seek its value at a given instant we could not thus proceed. If we measure the space described in the second which commences with the given instant would give us too high a value since the velocity is constantly increasing during that second. To measure the space described on any less duration from the given and compare it with that duration would give a *nearer* result because there would be *less* variation of velocity – but not an accurate result because there would still be *some* such variation. However, two things are clear, 1st., That by taking a duration sufficiently small we may (supposing our measures accurate) arrive as nearly as we please to the measure of the velocity sought, 2nd. That we can never by this process actually reach it. In these two particulars is involved the true idea of a Limit

---

* Quantity in the mere arithmetical view may be considered as extending from 0 nothing up toward infinity. In algebra we recognize it as extending also in an opposite direction towards negative infinity. Thus if we imagine a straight line infinite in both directions and fix upon a given point in it as a zero point or as the origin of measures of distance then distances measured from it in the one direction may be called positive and those measured in the opposite direction negative and the numbers by which they are represented may in the one case be affected with the sign + and in the other with the sign –.

which may be defined as a fixed value toward which some varying value may be made to approach as nearly as we please, but which it cannot be made to reach. A rate is a limit of a ratio. In the above example we have the following elements, 1st. Time increasing uniformly. 2ndly. Space described increasing with the time. 3rdly. Velocity which is the *limit* of the ratio which space described bears to the time of its description reckoning from a given instant.

Now the Diff[erential] Calculus takes up and solves the general problem of which the above is a particular example. Given the expressions of two quantities one of which X increases uniformly and the other increases with it but not necessarily in a uniform manner – required the expression of the limit of the ratio which the increase of the latter bears to the increase [A90.15] of the former reckoning from a given value of the former. By the expression of the limit of the ratio I mean an expression from which *exact* value of the limit of the ratio may in any particular instance be calculated.

That quantity whose variation is uniform as in the above case the element of Time, is called the independent variable. That quantity whose variation is referred to the variation of the former is said to be a *function* of it. The Differential calculus enables us in every case to pass from the function to the limit. This it does by a certain Operation. But in the very Idea of an Operation is involved as we have seen the idea of an inverse operation. To effect that inverse operation in the present instance is the business of the Int[egral] Calculus. This it accomplishes in a large number of cases but not in all. The process is called Integration.

The integral calculus proposes and in many cases resolves a yet higher problem. Given any relations among limits to determine the corresponding relations among the functions. This process is termed Solution of Diff[erential] Equations.

The integral calculus is the great instrument of physicomathematical investigation. The phenomena of the material Universe so far as they depend on figure and motion are usually of that kind which is characterized by continuous variations and they are moreover consequences of the laws of mechanics. The direct knowledge which those laws afford to us is a knowledge of the rate of change which is going on at a given instant. It is therefore a knowledge of limits. To estimate the total effect of such changes continued through a given time would require a knowledge of the functions from which those [A90.16] limits are derived and that knowledge can only be obtained through the agency of the Integral Calculus. Hence the extreme importance of that Science[9] as an instrument both for the verification of known laws and for the discovery of others.

Of the logical forms of reasoning which are exhibited almost exclusively in the mathematics the following are the most deserving of notice.

It is sometimes required to prove a theorem which shall be true whenever a certain quantity n which it involves shall be an integer or whole number and the method of proof is usually of the following kind.

1st. The theorem is proved to be true when $n = 1$. *2ndly.* It is proved that if the theorem is true when n is a given whole number, it will be true if n is the next greater integer. Hence the theorem is true universally. For according to the second principle established, if it is true when n is 1 it will be true when n is 2 and if it is true when n is 2 it is true when n is 3 and so on *ad infinitum.* This species of argument may be termed a continued *sorites.*

The real nature of mathematical analysis

That Geometry is nothing more than an application of ordinary reasoning to the particular subjects of which it treats is sufficiently manifest. In Algebra however and in all those applications of analysis in which symbols are made use of, another element is introduced the nature of which it remains to investigate.

We have seen that our conceptions of number and magnitude are subject to certain fundamental and necessary laws. If we represent those conceptions by symbols our symbols must be made subject to the same laws. In other words our language must be [A90.17] so constructed as to represent in its laws the laws of our conceptions as well as in its symbols the conceptions themselves.

Further we represent by the signs of addition, multiplication, etc., those operations to which the things we conceive of are subject, and by the sign of equality the relations which they may be supposed to bear to each other. This being done we possess a language which is in every respect an image of thought, the subjects of thought are represented by its symbols, the operations of thought by its signs and the laws of thought by its laws. All the operations of that language will consist in the substitution of expressions for equivalent expressions, the equivalence being determined by the laws above referred to. Grant that the data of a given problem are accurately represented by that language. Whatever conclusion then its lawful operations may conduct us to, will though merely formal in itself if it admit of interpretation represent a conclusion true in the subject. I say if it admit of interpretation because it is not true in thought that every operation can be performed upon every subject but it is possible in language to represent any operation as performed upon any subject, the laws which that operation obeys when possible in thought being still observed.

We may distinguish in any given example of symbolic reasoning the following elements.

1st. The data of the question considered as relations among things. These we may term the real premises.

2nd. The symbolical equations by which those relations are expressed. These we may call the symbolic premises. [A90.18]

Chapter III

[Extracts from a notebook]

—————————————[Probably between 1849 and 1850]—————————————

Miscellaneous Observations on Logic

It would perhaps be better to exclude the Baconian induction from formal logic & to call the "perfect induction" as the "proof by enumeration". It indeed strictly belongs to the deductive logic by which I mean that logic which deduces certain conclusions from premises.

The Baconian induction may be regarded as leading to material consequences — the common logic is formal [N.1]

*　　*　　*

The ideas which are suggested by the purposes of thought are the ideas of Succession of Distribution of Equality. The *subjects* of thought are *things*, or states of things or in other words *things* or *conditions*. We contemplate either objects in which case we assert relations existing among [N.7v] them or we contemplate the cases in which such conditions are fulfilled and affirm so to speak relations among those relations. The former case is that of categorical assertion the latter that of hypothetical or disjunctive assertion. [N.8]

*　　*　　*

The laws $x(y+z)=xy+yz$ and $xy=yx$ are the only laws of algebraic symbols devoid from the conception of quantity. The law $x^m x^n = x^{m+n}$ merely implies that if the results of the operation x be performed n times, and then m times upon the result, the final result is the same as if it were employed m+n times. In full it merely expresses a law of *number*, i.e. of the number which expresses how often an operation is performed and indicates that that number may be considered as a *whole* m+n, and resolved into parts. Hence the law $x^m x^n = x^{m+n}$ is a law quite independent of the nature of the operation denoted by x. It is for example true in the Differential Calculus that $(\frac{d}{dx})^m (\frac{d}{dx})^n = (\frac{d}{dx})^{m+n}$.

To the above[1] it would seem that we ought to add the law whose expression is

$$x+y=y+x. \qquad\qquad [\text{N.8v}]$$

*   *   *

Equal inverse operations of division give indefinite results which are equal on the proper determination of the indefinite function. [N.9]

*   *   *

If we had a problem to solve relative to some particular numbers such as 10, 20, 30, and could not conceive of others – we might investigate the laws to which 10 20 30 are subject and employ general symbols subject to these laws finally interpreting the result – the intermediate steps being uninterpretable *for these numbers*. We interpolate for other values of the general symbols. [N.10v]

In fact we may employ signs to represent particular conceptions determine [how] the laws of those signs operate in accordance with these laws so as to pass through forms quite uninterpretable with reference to the original conception but finally obtaining a result which is thus interpretable and rely[ing] upon that interpretation as correct.

Here the dominion of the laws is more general than the particular operations in which they are first observed to prevail. [N.11]

*   *   *

In applying the method of indeterminate multipliers in the calculus of Logic we pass through uninterpretable forms to forms finally interpretable through development. Number is introduced in the Constant multiplier. [N.11v]

There is an important distinction between the interpretation of the symbols in any proposed form of analysis and the forms under which these symbols enter into the analytical expression. There seems to be no limit theoretically to the former but the latter are apparently fixed and are the same for all symbols.

Suppose x, y, z to represent any operations we can then only combine them mentally by distribution or succession, and even some of these modes may give uninterpretable results. We can [N.11v] conceive of such forms as

$$(x+y)1 \qquad\qquad xy1 \qquad\qquad xx$$

and can in various instances interpret them and determine the laws of combination. But the very laws of combination have reference to special forms. An operation like an event is hypothetical, may be regarded either as producing its effect coordinately with some other operation — or in a determinate succession with reference to that operation or with reference to itself by repetition.

From this repetition we got

$$xx \text{ or } x^2, \qquad\qquad xxx \text{ or } x^3 \qquad\qquad \&c$$

whence we have the idea of number. Or we have

$$x+x = 2x, \qquad\qquad x+x+x=3x \qquad\qquad \&c$$

whence also the idea of number. In the latter case it must be supposed that the $x1$ in

$$x1+x1+x1+ \qquad\qquad \&c$$

refer to different or mutually exclusive entities so that we may have the possibility of aggregation.

We can also conceive of the existence and investigate the nature of an operation which being performed n times may produce the same effect as the operation x performed [N.12] m times and thus we get $x^{m/n}$. And we can conceive of an operation the result of which being aggregated n times shall give the same result as the operation x aggregated m times, i.e. of $\frac{m}{n}x$ and from these considered as single operations we have by the principle of aggregation already stated

$$ax^\alpha+bx^\beta+cx^\gamma,$$

abc $\alpha\beta\gamma$ being integral or fractional.

It is clear from the above that the idea of number is introduced by the repetition of an operation as in successive periods of duration. It is equally introduced by the consideration of individuals of the same kind as defined by a common name simultaneously existing.

Both the system of elective symbols and the system of numerical magnitude set out from the same [N.12v] point – the consideration of the whole or unity. In the case of magnitude we are led to contemplate different wholes without speculating on the constitution of any one – in the system of elective symbols we have but a single whole, the Universe whether it be the actual Universe or some portion of it to which the discourse is limited and we define more or less its constitution by specifying the classes which enter into it.

A great distinction between the system of elective symbols and that of quantity is that in the former the equation may relate to several distinct classes of things included in the universe 1 – in the former [latter] we only relate to one description of thing viz. the unit itself to which all the numerical coefficients are supposed to apply.

In the mathematics of quantity we proceed by the combination or repetition of the unit. In the mathematics of Logic we proceed by the analysis of the Unit.

The analysis of the Unit is effected by the combination or repetition of operations and hence regarding [N.13] one of these operations as a whole we have quantity introduced into the elective equation.

The language of quantitative mathematics is independent of the particular nature of the unit. The language of analysis in general is therefore independent in its forms of the particular nature of the operation represented by the operating symbol.

A good example of the real nature of an elective symbol is afforded by the discontinuous integrals which are employed as multipliers in the calculation of the values of definite multiple integrals according to the method of M. Dirichlet.[2] For these multipliers are equal to unity for certain points in space & to 0 for others. The product of two such multipliers would therefore represent all the points in space which are common to the representations of each. [N.13v]

Of Inverse Operations

It has been remarked that beside considering the results of direct operations we can propose to determine the nature of a subject upon which the performance of a given known operation shall produce a given known result.

It would appear that question is in its very nature indefinite. If we obtain a solution we cannot a priori prove that it is the only one — only by considerations proper to the cases. It is like the inquiry into *cause*. If a particular state of things as cause produces a particular effect we cannot from this alone be sure that there exists no other state of things which will produce the same effect. [N.14]

*   *   *

Introduction

Relation of this work to the Mathematical Analysis of Logic exhibited in the history of my investigations. Its principles the same. But that work founded upon the plan of the common Logics. Its insufficiency for the examination of complex arguments. The general inutility of technical forms. Attempt to arrive at a general method upon the principles previously arrived at. The present work the result of that inquiry.

When completed this work will consist of two parts the science of necessary and the science of probable inference. The present work confined to the former subject which serves as an essential preliminary to the latter.

Analysis of the work according to its chapters

1st The nature of signs in general. The analysis of the elementary signs of reasoning with their laws.

2nd The division of Propositions into primary and secondary – the expression of the former. [N.15]

3rd The Rules of Elimination Solution and Development.

4th Their application 1st to the analysis of definition 2nd to general reasoning.

The expression of Secondary Propositions. Their analogy in rule and expression with primary propositions. Nature of the inferences to which they lead.

Uses of this inquiry in a speculative point of view.

As throwing light on the laws of the human mind, on the question on Human Liberty etc. As illustrating the progress of ancient philosophy and metaphysics etc.

Uses in a practical theory of view.

As affording a general method of the correctness of any process of reasoning.

Hence its utility extends to all sciences which proceed from exact definitions and general principles. Conclusions are drawn by common reasoning. [N.15, verso]

Political Economy Theology Natural History Law and Morals? are examples of this kind.

In fact there are two processes in every true science. One is the discovery of true principles, the other the deduction from them of correct conclusions. The former the more difficult and more important task. The latter however equally *necessary*. Its importance to be judged of by this consideration. Suppose that we could for ever banish false vague and inconclusive reasoning from the discussion of social moral and other questions. Would not those sciences receive a vast benefit [?] Would it not become the main business of readers to inquire into the truth of the premises and when it was seen how small is the number of these general principles (out of the circle of the confessedly exact sciences) upon which an absolute dependence can be placed, it might tend somewhat to promote peace and charity in the world. To this end it would contribute much if men considered that while probabilities are [N.16] in most instances a sufficient guide of conduct and of action – they are not sufficient to warrant us in uncharitable condemnation of those who differ from us. [N.16, verso]

*   *   *

Of signs[3]

1. Language is not only a medium for the expression of Thought but also an instrument of the Reason.

2. Its elements are signs subject to laws.

3. A sign is an arbitrary mark having a fixed interpretation & susceptible of combination [N.17] with other signs in subjection to definite laws.

4. 1st Arbitrary. Therefore a sign may be either a notion or a word, or a letter. 2nd fixed. Therefore it must always be used in the same sense throughout the discourse. 3rd susceptible of combination according to definite laws. A combination of signs may be replaced by a single sign, or ...[4]. Signs are thus susceptible of combination in so many ways that it becomes important to inquire where are really those elements and signs of Language which are essential to & sufficient for all the purposes of Reasoning. They are

5. 1st Signs expressive of a name quality or circumstance. These have contraries expressed by *not* and are subject to the commutative law.

6. 2nd Signs which relate to the mental operation of dividing a whole into its parts or aggregating parts into a whole.

Subject to the law of distribution.

7. 3rd Signs expressive of the substantive & propositional relation implied by *is* or *are*.

8. Law of transposition. [N.17v]

9. More [?] particular examination of the analogy between the literal symbols above and the adjective.

The adjective in common discourse elective as well as attributive. Sole difference that the literal symbol when standing bare or alone becomes a substantive or that the word *things* is always understood, those things being confined to the Universe of discourse. [on N.18]

10. Other signs are either subsidiary or resolvable into the above or they have an analogous meaning and are subject to the same law.

11. Thus the prepositions are subsidiary to the purpose of expressing quality or circumstance. Verbs are resolvable into the substantive verb & an element expressive of circumstance. The conjunctions if, either, or &c are used in the expression of relations among propositions and their relations. It will in the end be seen as analogous to those which have been considered as expressible by signs subject to the same laws.

12. The laws above assigned for the fundamental signs of reasoning are identical with the laws of the signs of quantity in Algebra. [N.18]

Part B

The Philosophical Interpretation of a Theory of Logic

Chapter IV

Prolegomena

————————————[Later than 1854]————————————

1. Writers on Logic may be divided into two great classes as the first consisting of those who regard all Logic as having no other *ultimate* object than Reasoning and who contemplate all reasoning as typified in the Syllogism, the second consisting of those who recognize the *independent* value of the study of the intellectual operations of Conception and Judgment and who endeavour to refer the Laws of Universal Logic including therein the syllogism to certain primary laws as principles of thought e.g. the principles of identity, contradiction excluded middle and sufficient reason.

2. The consent which binds together the members of the first of the above classes is tolerably complete. The theory of the all-sufficiency of the syllogism is a principle of union which has the great advantage of being easily intelligible and of being *one*. It does not harass the mind with conflicting claims.

3. The consent however which exists among the members of the latter of the above classes is very imperfect. Some make no mention of the principle of sufficient reason, but regard all Logic as based on the principles of identity, contradiction and excluded middle. Some again regard the principles of identity and contradiction as one. Other differences exist e.g. as to the import of the proposition.

4. As the principles of identity and excluded middle [A94.1] in which at least the members of the second class agree are true principles and are moreover either *first* principles or as consciousness informs us lie close to the very ground of the possibility of thought (it matters not here that they have been regarded by some as metaphysical and not logical and by others have been considered frivolous and trifling). A theory of Logic which does not both recognise them and give to them a very important place must be a very imperfect theory. It does not suffice that their existence should be casually noticed.

5. On the other hand it is very evident that the principles of identity, contradiction and excluded middle are rather principles which relate to conception and to judgment – to the power by which [we] conceive of things as existing and as existing in relation expressible by propositions – than to reasoning. To the latter they seem so belong only or chiefly in an *implicit* manner inasmuch as reasoning presupposes conception and judgment as the sources from which the materials upon which it operates are derived.

6. Even if to the above mentioned principles of identity contradiction and excluded middle we add the principle of sufficient reason we are still far from possessing an adequate basis for the Science of Logic. For to say that no proposition must be accepted as true no conclusion admitted as valid except upon sufficient grounds (and such is the form in which the connection of the principle of sufficient reason with Logic is usually expressed – indeed I do not know of any other form in which it has ever been expressed) affords no satisfaction to the enquirer who demands what are the grounds in question. The principle of sufficient reason has no more *special* relation [A94.2] to the laws of inference than has the principle of universal causation to the laws of any of the physical sciences. It furnishes the ground upon which we may affirm our right to enquire what the laws of logical inference are – it does not anticipate or forestate the results of that enquiry.

7. Accordingly it may be remarked that they who perceiving the insufficiency of the principles of contradiction, identity and excluded middle have proposed to supply the defect by the principle of sufficient reason have never shown or even attempted to show how the *acknowledged* rules of syllogism are to be deduced from that principle. Still less have they attempted to evolve out of that principle whether simply or in combination with the other principles above referred to any system of general rules and methods applicable to all logical inference whatever.

8. Nor again does it appear that even if we make abstraction of all those laws of thought which have a *primary* reference to conception and judgment and confine ourselves as far as possible to the consideration of the process of inference alone that any one has shewn [that] a sufficient and complete account of that process is contained in the laws of syllogism. It has indeed been affirmed with endless iteration that syllogism is the one universal type of all valid inference, but no proof of that assertion has been offered. The difficulty of reconciling with that assertion the cases of what is termed "immediate inference" has been felt and more extended examination would have shewn that this is but one of the difficulties with which the theory of the sufficiency of the syllogism has [A94.3] to contend.

9. Although logic when reviewed in relation to any other science occupies a position of priority and of predominance we are not exempted in the study of it from those conditions to which we are subject in all other investigations, which have truth for their object. Facts of external observation are indeed replaced by facts of consciousness but we are still bound to attend to them as

facts. We must not frame theories which either *contradict* them or *neglect* them. For instance the enquiry into the distinct operations of thought is entirely based upon an appeal to consciousness. We could never have affirmed *a priori* that thought is resolvable into the three operations of Conception, Judgment and Reasoning. But having established this fact by the direct testimony of consciousness as manifested in individual acts of thought, we must henceforth respect it as a *fact*. So too if further analysis of the facts of consciousness should show that any of the above operations as Conception is resolvable into other operations distinct in species but generically agreeing we are bound to respect that analysis and the facts to which it leads in our theories and systems.

10. And it concerns us not only to analyze and to classify the operations of thought, but also to determine their laws. The former object howsoever completely attained can only constitute the Natural History of Thought – the latter is indispensable to its Philosophy.

11. As every law not only colligates but arranges many facts, so does every theory which has any real foundation in [A94.4] the nature of things bring into some harmonious connection and subordination many laws. All science tends toward unity. Were it otherwise we should be lost in the multiplicity even of laws and relations. And thus in the science of the intellectual operations where while some recognise only the one principle of the dictum of Aristotle as governing every form of inference others have discovered in each distinct form of argument the embodiment of a distinct principle there is at least the presumption of analogy that some system of primary laws not less truly partaking of the character of unity than the one, and wider in the compass of its relations than the others may exist.

12. But dismissing all considerations of analogy we possess a negative test of the completeness of any account of laws of thought professedly based upon the testimony of consciousness in the completeness of the system of methods and processes into which the supposed laws are capable of being developed. For if there exist any valid processes of logical inference which are not included in the list of their consequences (since methods and processes are truly the consequences of laws and do not spring up arbitrarily into existence) we may certainly conclude that the system of laws in question is imperfect.

13. And on the very same ground *viz*.: that methods and processes do not spring up arbitrarily into existence there is a strong presumption that where a system of methods is as such complete i.e. where it accomplishes all the objects which can be demanded of it, there is I say in such case a *presumption* that the basis upon which it rests is not merely empirical or analogical. [A94.5]

14. The conclusion to which these observations tend is that in the attempt to construct the Science of Logic we must seek 1st. to distinguish the several operations of thought and determine their laws, consciousness here

taking the place of experience. 2ndly. to construct upon the basis thus obtained the *general* methods and processes of the science. 3rdly. To ascertain whether there exist any valid processes of thought (consciousness being again the seat of appeal as to their validity) which are not included under the general methods and processes above referred to. [A 94.6]

Results of the Analysis of Conception

1. The analysis of Conception may be resolved into two distinct portions: 1st. That which relates to the primary genesis and to the nature of a conception e.g. to the mode in which from our experience of individual men we arrive at the concept expressed by the term "men" and to the import of that term 2ndly, That which relates to the combination of concepts already existing as elements of thought e.g. to the operation by which from the given concepts gold and mountain we form the concept golden mountain.

2. It may be doubted whether the first of the above processes viz. the primary genesis of concepts from the presentations of the senses or of experience falls within the domain of formal Logic. The question is in some degree parallel to the question whether the origin of our conceptions of Number falls within the province of Arithmetic as a science. Perhaps the proper answer would be that the enquiry into the genesis alike of those concepts which are expressed by general terms in Logic and of the concepts of Number in Arithmetic belongs to an earlier stage of psychology than either Logic or Arithmetic but [are] so allied with those which follow it that its neglect must give a character not indeed of imperfection but of isolation and also of technical formality to the study of them.

3. All that need here be said of the genesis of logical concepts is that they originate in certain prior acts of the understanding following in an orderly sequence viz.: 1st. the direct presentation of the senses or of experience 2ndly. comparison by which we note the particulars in which different [A 94.7] individuals agree or differ 3rdly, generalization by which those which agree in certain selected particulars are contemplated by the mind under a certain mental type common to them all and which mental type may be represented by an external sign – a notion – a word – a symbol. The mental type thus capable of association with an external sign is what is termed in Logic a concept – it is that which corresponds within to the sign without. It is to be observed that a concept is not an abstract idea in the sense in which that term is commonly employed. It does not represent an abstract or ideal entity possessing only the essential qualities of the class of things represented to the exclusion of all other qualities. In its most definite form it is the mental representation of a particular individual but with the convention that it is to stand for *any* individual of the class. It is a *specimen* understood as a *type*. It is not however requisite that it should resemble an

individual at all. That which is essential to the office of the concept in thought is that and only that which is essential to the sign in language, *viz.* that it should have a fixed association (fixed only however by a convention of the understanding) with the individuals for which it stands, and should in virtue of that association represent them in thought.

4. As to the second part of the analysis of conception viz. that which relates to the combination of conceptions already given the following results are very generally accepted viz.: 1st. *that we are able to conceive of things under two wholly distinct relations – the relation of whole and part and the relation of subject and quality and that these relations are the basis of the two great logical processes connected with* [A94.8] *conception viz. division and definition.* The relation of whole and part is exemplified when we contemplate any collection of things as formed of parts which by simple aggregation or putting together form that whole just as we may conceive of the sun, the primary, the secondary planets etc., as comprising the solar system omitting from our idea of the solar system any notion [of the] physical relations of those parts.

The relation of subject and quality is exemplified when we contemplate any subject of thought as possessing different qualities by any one of which it might be *described* but by the combination alone of which it can be *defined.* e.g. Man may be described as *rational* or as an *animal* but he is defined as rational animal.

It is very manifest that the mode in which the conceptions implied by the terms *rational* and *animal* are combined in thought in order to form the concept *man* is different in *kind* from that by which suns, planets etc., are put together to form the concept *solar system.* And this difference is recognised in the construction of language. We usually employ the conjunction *and* to express that aggregation by which parts are simply collected into a whole and thus we say the sun *and* the planets etc., are the solar system. No conjunction is employed or if one is employed it is only employed by way of analogy to connect together concepts which stand in the relation of subject and quality. We say Rational animal – not rational and animal – is man.

The analogical employment of the conjunction is exhibited in such expressions as "wise and good man". That it is only analogical will appear from the fact that we may omit the [A94.9] conjunction without impairing the sense. "Wise good man" is the same as "Wise and good man". The And is in fact wholly superfluous.

As to the laws of division and definition it is posited

1st. That the parts of a division must be mutually exclusive and must together make up the whole.

2ndly. That in definition we cannot combine together a concept and its contradictory e.g. we cannot apply the epithets rational and not-rational to the same individual. This is the formal statement of the principle of contradiction.

Thus, to recapitulate what has been said, the received analysis of Conception consists of two parts. The first part explains in a manner to which there is not much to be added the *genesis* of conceptions from the presentations of the senses. The second part recognises in conceptions actually formed the two relations of whole and part of subject and quality. The former relation it exemplifies in the process of division which it contemplates as subject to two laws viz.: 1st. that the parts must be exclusive 2ndly. that they must together constitute the whole. The latter relation it exemplifies in the process of definition which it contemplates as subject only to the "principle of contradiction".

The above is a sufficient summary of logical doctrines on the subject of conception – for although the principle of excluded middle is sometimes regarded as regulating that process it belongs properly speaking to the analysis of judgment.

The first observation which presents itself in considering the above scheme is that neither of the processes of division and definition as above described is necessarily a simple [A94.10] process. Each is composed of a more elementary character. Thus division involves an act of judgment in addition to acts of conception. To conceive of a whole belongs to conception, to conceive of parts aggregated together also belongs to the faculty of conception, but to realise mentally the *equivalence* of these two products of conception belongs to judgment. Its outward expression is a proposition e.g. the sun and the planets *are* the solar system. Similar observations may be applied to *definition*.

Accepting however that limitation which applies the term division to the operation by which we can see of parts aggregated together even this operation is generally resolvable into distinct steps or stages each consisting in the addition of one new part to the aggregate of those which have been connected before. When we say "the sun, the planets and the comets" the actual processes which take place in the mind of a person to whom the language is addressed are 1st He forms the conception of the sun, 2ndly, he connects with this concept that of the planets so as to form the whole expressed by sun and planets 3rdly, with the whole thus formed he connects the concept comets so as to form the larger whole "sun planets and comets". Here beside that primary operation by which a sign calls up in the mind the conception of the things signified we have two other operations each of which consists in the connection or to speak by way of analogy the *aggregation* of two concepts in thought so as to form a whole. We cannot resolve the latter operation into anything more elementary.

This operation of aggregating parts into a whole suggests another operation which is properly its *inverse* viz. the [A94.11] operation by which from the concept of a whole we by removal or separation of the concept of a component part arrive at the conception of the remaining part.

It is plain that the operation just described is *inverse* to the one before considered. As from the conception of [the] sun, we by addition if we may so speak of the conception planets arrive at the aggregate conception solar system so from that aggregate conception solar system by *"subtracting"* if we may so speak the conception planets arrive at the conception "sun". Here the second operation exactly undoes what the former operation has done and brings the mind back to that state from which it set out. And this is the true idea and definition of an inverse operation.

Although the inverse operation just described is suggested by the prior operation of addition to which it stands in the relation of an inverse, its existence does not *depend* upon that association. It is intelligible in *itself*. We can directly perform that operation which consists in the removal or subtraction of the conception of a *part* from the conception of the whole in which that part is involved. And it may be added that the symbol or verbal equivalent of this operation forms an element of ordinary speech. The word *except* when employed as a *preposition*[1] not as an adverbial conjunction expresses the operation in question.

Again the process of definition is not usually a simple process. Setting aside the act of judgment which it involves it consists of one or more distinct stages or steps each of which involves a mental operation. If we define "man" as *animal rationale* then beside that operation of judgment which [A 94.12] pronounces the equivalence of the two concepts "men" and "animal rationale" the formation of the concept "animal rationale" implies a selecting out of the class *"animal"* all those individuals which possess the attribute *"rationale"* and fixing the attention upon those selected. If the male individual of the race should be defined as *animal rationale barbatum*, in the formation of this complex concept two distinct operations of the above character will be involved viz.: 1st. that by which from the class animal we select and fix the attention upon the individuals possessing the quality *rationale*. 2ndly. that by which from the class whose concept is thus formed we select those individuals which possess the further quality *barbatum*. And such processes may in the more complex examples of definition be indefinitely repeated.

The operation which has just been described is not represented like the two former ones by a distinct symbol or verbal equivalent in language – but it is represented by certain well-marked constructions – e.g. the *immediate* connection of the adjective with the substantive – the immediate connection of several adjectives which are applied to the same substantive, no sign of connection being required – the adoption of some periphrases involving the relative pronoun, where the concepts to be connected are separately expressed by substantives e.g. those minerals which are of vegetable origin and possess commercial value. – Here three concepts are combined not by way of the aggregation of parts into a whole but by that species of composition which enables us to rend more definite and complete the conception of a subject by distinct acts each of which involves the attribution of a quality or property. [A94.13]

To the direct operation which has just been described there exists a corresponding inverse which it also serves to suggest. We may define that inverse as the operation by which from the conception of a given class of things we ascend to the conception of some larger class from which the given class would be formed by the mental selection of those individuals which possess a given property. Thus as by a direct operation of thought we *descend* from the conception animals to that of rational animals by the mental selection of those individuals which possess the property of rationality – so by a corresponding inverse operation we should ascend from the conception of rational animals to that of animals.

It will at first sight seem that the inverse operation just described is the same as the mental operation commonly known by the term "Abstraction" – and the example which has just been given of it is in reality a case of Abstraction. But in strict truth Abstraction is only a particular form of the inverse operation in question when considered in its largest sense. What for example is the most general conception which we can form of that class of things from which if we select all which possess the property of rationality the result attained will be the entire class rational animal. It is evident in the first place that the class sought must include the class rational animals in the second place that it must contain no rational beings besides. No other condition can be assigned. We may therefore say that the class from which by selecting rational beings we obtain rational animals is in its most general conception a class which comprehends all rational animals with an indefinite remainder of irrational beings. [A94.14] This general conception includes the particular conception of "animals" before obtained. For the class animals consists of all rational animals together with a definite remainder of irrational beings viz. irrational animals. The indefinite remainder which enters into the general conception is here made definite and specific. And thus *abstraction* in its ordinary sense is seen to be a specific determination of the inverse operation whose larger meaning has been exemplified.

In that larger meaning the operation under consideration has no verbal symbol or equivalent construction in language. It is in fact only conceivable by means of that operation of which it is the inverse and therefore only expressible by an *indirect* definition. We can only define it as the undoing of something the doing of which can be defined *directly*. Further it may be observed as a consequence of what has been now said that the inverse operation does not enter among the elements of definition or description as the three other operations before considered do. In its more general form it is capable of fulfilling this office because of its indefinite character. In its more specific form of abstraction its employment even supposing the nature of the specification to be thoroughly understood would be superfluous. We should never think of employing instead of the term "animals" the periphrasis "that class of beings from which if we mentally select those which are rational we shall obtain the class rational animals."

But although the inverse operation under consideration does not form any part of definition either in the special sense in which the term definition is usually understood or in the larger sense explained in ( ) it is a *real* and a [A94.15] very *important* operation. If it does not belong to the faculty of conception as exercised in *definition* it does belong to that faculty as exercised in some of the processes of inference. The following would be admitted to be a legitimate problem in Logic. Given the definition of clean beasts in Jewish Law viz.: Clean beasts are beasts which both chew the cud and divide the hoof required a *direct* expression for beasts which chew the cud in terms of clean beasts and beasts which divide the hoof. Now the first step to the solution of this question obviously is to say "Beasts which chew the cud must be a class of beasts from which if we select those which divide the hoof² we shall obtain the class of clean beasts". But this only indicates that the complete solution depends upon the performance of the inverse operation above described and here more than one question of deep interest and significance presents itself. The operation which we are called upon to perform is the inverse of a direct operation which is undoubtedly a simple one. But is the inverse operation itself a simple one? If it is wherefore has it no verbal symbol no equivalent construction in language. If it is not what are really the mental operations of which it is composed? What new mental processes must we take into our account? These are questions of which it is needless to say the ordinary treatises of logic as they contain no intimation so they offer no solution.

Something must here be said of the origin of *negatives* although their theory can only be completed when we come to treat of propositions and of the relations to which they are subject. Indeed we must to some degree anticipate that subject here. Such a proposition as Men are not perfect may be [A94.16] understood in two distinct ways – distinct but so connected that the one necessarily implies the other. 1st. It may be considered as a *denial* – as if some assertion had been made that men were perfect or that some of the race [were] perfect and as if the truth of that assertion were to be denied. A denial must be a denial of the truth of a proposition and there is a branch of Logic as will be seen which relates to propositions in their special attributes of truth and falsehood and in the relations flowing from those attributes. To that branch of Logic we shall hereafter have occasion to attend. 2ndly. It may be considered like an ordinary affirmative proposition as an *assertion* viz. as an assertion of which the subject is men and the import that they are not-perfect i.e. that they are imperfect beings. In this mode of considering the proposition things are divided into two great classes viz. perfect beings and imperfect beings and the class men is referred directly to the latter.

According to the former mode of consideration the *negative* particle *not* determines the *nature of the proposition* according to the latter it determines the *import of the predicate*. But both are valid and as has been said the one *implies* the other.

Inasmuch however as the second mode involves only those operations of thought which have at present been considered operations which belong directly to conception and do not presuppose the relations of *propositions* it is necessary to consider it first. Two methods of passing from the concept "perfect beings" to the concept "imperfect beings" present themselves. We may employ either the second or the fourth [A94.17] of the mental operations which have been described.

First we can do this by *subtracting* the positive concept of perfect beings from the positive concept of universe, the aggregate of all existing beings. What remains will be the concept imperfect beings.

In this mode of genesis we avail ourselves of the principle that any concept as perfect beings together with the corresponding privative imperfect beings will make up the universe that to every existing individual one of the two opposed terms perfect being, imperfect being may be applied (Law of excluded middle) and to none of them both (Law of Contradiction).

2ndly. We can also arrive at the concept imperfect beings by enquiring what class of beings that is, which combined by the third operation of thought with the concept perfect beings will produce the concept *nothing*. The answer is that the class must consist of imperfect beings. It need not be the whole of that class. Any indefinite portion of it satisfies the conditions. The class which by combination (operation 3rd.) with the class perfect beings produces the concept Nothing is in the most general manner defined by the expression Some imperfect beings.

According to this mode of genesis we make use of the principle of contradiction only. But it is to be observed that it does not as the previous modes did define the *extent* of the concept generated. It determines its quality but not its quantity.

The first mode which rests upon both principles of contradiction and excluded middle is then the only perfect mode [A94.18] of passing from the concept perfect beings to the concept imperfect beings without transgressing the range of the operations described in the previous sections.

To recapitulate what has been said. The negative *not* either directly qualifies propositions or it directly qualifies the terms of propositions. These two offices are so related that the one usually implies the other. The latter office alone however viz. the qualifying of the *terms* of propositions falls entirely within the sphere of conceptions as defined by its fundamental operations already described. Within that sphere the only definite mode of passing from the complete concept "perfect beings" to the complete concept "imperfect beings" is by subtracting (operation 2nd.) the concept "perfect being" from the concept universe.

It is to be remembered that when we say "within the sphere of conception" we mean thereby to *exclude* the introduction of the notion of a proposition which properly belongs to judgment. Undoubtedly the term imperfect beings does usually convey an idea of *denial*, and therefore of a *proposition*. But it is important to shew that the concept may by legitimate operations of thought be generated without that notion.

Thus far we have been occupied in classifying the operations of conception and examining their meaning. Another object of not less importance is to determine their laws.

These have been investigated in the Laws of Thought and the following results are chiefly taken from that work. But some important additions will be introduced chiefly relating to the conditions under which the operations are possible.

The operation of addition presupposes as a condition of [A94.19] its performance that the things added together should be distinct. When we say Men and Women it is implied by the very form of the expression that the concepts Men [and] Women belong to distinct classes of things.

It will be said that in a parallel example "orators and poets" no such implication as of the distinct and mutually exclusive character of the concepts involved is really made – that the expression is used and properly used with a distinct understanding that there may be orators who are poets – that the two faculties in question may be combined in the same individual. If however we attend to the mental process rather than to the language we shall perceive that when by the term "orators" the concept of the class thereby signified has been raised in the mind the additional words "and poets" do not lead us to attempt to annex to that concept by way of aggregation the entire concept poets but only the concept of such poets as are not already included in the class "orators". I rather apprehend that the real office of those words is to lead us first to resolve the class orators into two portions viz. the orators who are poets and the orators who are not poets and then to add to these concepts the new one of poets who are not orators. And here the final concepts actually added together satisfy the condition of being mutually exclusive. The use of language is not always determined by its mere *form*.

The condition of possibility of the operation of addition having been thus stated the first law of the operation is the following. *If two or more concepts are added together the order in which they are presented to the mind in the performance* [A94.20] *of the operation is indifferent.*

Thus if to the concept men we add the concept women the result is equivalent to the result which we should obtain by forming first the concept *"women"* and then adding thereto the concept *"men"*. The term *men and women* represents the same class of things as the term *women and men*; but the order of the mental processes which the one term expresses is different from that which the other term expresses.

When we endeavour to illustrate the laws of thought by the forms of language we must be careful to distinguish between the two offices or functions of language, viz. its office of *representation* by which words represent outward things, and its office of *expression* by which words express the states and operations of the mind. If this distinction is forgotten such illustrations as that which has just been given must appear frivolous. What it may be asked do we learn by being told that the expressions "women and men" [and] "men and women" are equivalent? Certainly we do not thereby arrive at the knowledge of any *material* truth; but we do learn that certain real operations of thought are subject to a certain law, which it may be observed does not belong to them merely as operations of thought but as operations of thought of a certain kind.

The condition of possibility of the second operation of conception viz. of subtraction is that the things to be subtracted should be entirely included in the things from which they are to be subtracted. Such an expression as "All men except negroes" is unintelligible unless it be supposed that negroes are men.

This condition being attended to the operation of [A94.21] subtraction is governed by a law similar to that which governs the operation of addition to which it is inverse viz.: *If different acts of subtraction are performed in succession the order in which they succeed each other is indifferent.*

Thus if from the concept men I first subtract that of negroes then that of Europeans the result will be the same as if from the race men we first subtract the Europeans and then the negroes.

We are hence naturally led to consider the case in which any series of acts of addition and subtraction follow in any mixed order. *It is evident that supposing the conditions of possibility still observed the order of the mixed operations in question is perfectly arbitrary.* If to the class trees we add mentally the class flowers and then mentally subtract the class pine trees the result is the same as if from the original class trees we first mentally subtracted the pine trees and to the concept then remaining added that of flowers.

But though the above is the most general expression of the law of addition and subtraction it admits of doubt whether it is a development of the simple law of addition first enunciated. For it is to be remembered that although the operation of subtraction is one the possibility of performing which independently of any direct and explicit reference to addition seems at first sight to be manifest yet it is capable of perfect definition as the inverse of addition. The analysis of consciousness seems further to indicate that in all cases the reference to addition though only implicit is real and indispensable. [A94.22]

On the Foundations of the Mathematical Theory of Logic and on the Philosophical Interpretation of Its Methods and Processes

——————————————— [1856] ———————————————

It may be necessary to inform the reader that the mathematical theory of which some illustrations will be given in the following Essay is the one which I have developed in a treatise published about two years ago and entitled "An Investigation of the Laws of Thought in which are founded the Mathematical Theories of Logic and Probabilities". It may also be proper to premise that I do not employ the term mathematical with reference to the theory of Logic as if with any covert implication that reasoning always and essentially involves the ideas of extension, magnitude, number etc. but that I employ that term solely as descriptive of the methods and forms on which the theory set forth in the work above referred to obtains its practical development. The foundations of that theory are laid professedly at least where alone the real foundations of Logic can be laid in an analysis of the intellectual operations. As moreover this paper is in some measure introductory to another containing a special application of its principles[1] I deem it right to state the

circumstances in which it originated. Some time after the publication of the
"Laws of Thought" my attention was directed by the Bishop of Edinburgh to
a question in the theory of probabilities not noticed in the above work upon
which conflicting opinions had been formed by different writers, and the
fallacious character of some of the reasonings which had been employed was
at the same time pointed out. In the course of the correspondence which
followed I was led to express a hope that Bishop Terrot would publish his
observations upon the subject and I placed at his disposal such results as I
had myself obtained. Without entering further into the details of the
correspondence let it suffice to say that it was finally agreed that the
publication of an analysis of the question in accordance with the principles
developed in the Laws of Thought should be undertaken by myself.[2] As this
analysis involved the application of the Mathematical Theory of Logic, I
found it necessary after an unsatisfactory attempt to combine that
application with a philosophical statement of its principles in a single paper
to divide the task. The present essay contains therefore an account freed as
far as possible from the language of symbols of the grounds upon which the
mathematical development of the Science of Logic rests. It is proposed in a
subsequent paper to publish the solution of the [B2.1] question in the theory
of probabilities to which reference has been made.[3] I am not without a hope
that the present attempt may possess some degree of independent interest to
those who are interested in the Philosophy of Mind. The need of some such
exposition of the principles of the Mathematical Theory of Logic as should be
intelligible to those who had little familiarity with the use of symbols has
more than once been urged upon me. Even among those who have
approached that theory with all the advantages which previous mental
discipline could furnish some need of further exposition has been felt. This
may be due in part to the novelty of the theory but it is due I believe in a
much larger degree to the imperfect mode in which it has been presented as
a philosophical system in the Laws of Thought. This defect I trust to be able
in some degree to remedy here. I shall endeavour to state first those general
principles of the philosophy of Logic which are anterior to and superior to all
the special forms of its development. I shall then investigate in a manner
which will I trust be deemed far more complete than my former efforts the
primary laws of the faculties of Conception Judgment and Reasoning and
briefly state the grounds furnished by the investigation for the mathematical
Theory of Logic. I shall finally shew and this part of my design is I believe
new that even the methods developed through the medium of the
mathematical theory admit of illustration in ordinary language and may be
understood and to some extent applied by those who are unacquainted with
the use of mathematical symbols.

    As a consistent and well understood nomenclature is of essential
importance in all such inquiries as the present it may be proper to say that
by the terms Conception Judgment and Reasoning I shall in accordance with

the usual practice of logicians designate both the intellectual *faculties* to which these terms are usually understood to have reference and the *operations* of those faculties. Wherever there is danger of ambiguity from this source, I shall explain whether it is in the sense of *faculty* or *operation* that the particular term in question is employed. The term Conception is commonly used in a third sense viz. to designate the *product* of the faculty of operation which that term represents. This seems objectionable and the substitution of the term concept suggested originally I believe by Sir W. Hamilton has recently obtained ground. The use of the term *judgment* to represent a *proposition* although sanctioned by high authorities appears to be liable to an objection of the same kind. In the present essay the products of the respective operations of Conception Judgment and Reasoning will be represented by the corresponding terms Concept Proposition and Conclusion or Inference. The word *Inference* I of course employ here in the sense of "inferred proposition." When it is necessary to speak of the intellectual faculties or operations generically, I shall employ such terms as "Thought" the Rational Faculty etc. Nor does this recognition of a higher unity merely depend upon classification or upon our internal consciousness of the oneness of the understanding. Between the intellectual operations of Conception Judgment and Reasoning exist clear scientific distinctions but the investigation of their actual laws demonstrates also the existence of a basis of unity. Thus as will hereafter [B2.2] be shewn the laws of Conception are expressed in the form of necessary propositions and hence their recognition involves an art of Judgment. On the other hand the foundation of all necessary propositions is to be sought for in the laws of Conception.

That which we conceive I shall term when considered in itself the *object* of thought. As any object may be presented to us in different aspects we may form of the same object different concepts. Language as the instrument of thought is related both to the object and to the concept which we form of it. The former as external it may be said to *represent*, the latter as internal to express. In the following pages frequent occasion will arise for referring to the functions of language and the above distinction will be kept in view. Indeed the strict etymological use of words is often of itself a safeguard from errors against scientific precision. An unconscious philosophy finding expression in the common sentiment and approval of mankind has consecrated terms physical in their origin to intellectual and moral uses and this it has done not arbitrarily but in accordance with a real analogy connecting the material and the mental world.

The tripartite division of the intellectual operations into Conception (more commonly termed Apprehension), Judgment and Reasoning has been recognized in all ages and it has been agreed by all who regard Logic as a science that it is the business of that science to investigate the laws of those faculties. Perhaps it has not been so generally perceived that these laws may in some measure depend upon the nature of the object of thought or more strictly speaking upon the nature of the general notion under which that object is contemplated. The concepts with which we have to do in Arithmetic and Geometry are peculiar and for aught that can be seen beforehand this peculiarity may affect nay it actually does affect the laws of the intellectual operations within the limits of each special province of thought. Still it is true that there exists above all these differences a philosophy of the intellectual operations which furnishes equally to the formal logician and to the student of the relations of Number and Magnitude a basis of first principles. There is indeed a large sense of the term Logic which embraces this general philosophy of the intellect. And such a sense it must be admitted accords well with the derivation of the term – ἡ λογική,[4] the science which is concerned with the operations of the rational faculty in general. The science of Formal Logic however if estimated not by the claims which have been set up for it but by the conquest which it has actually achieved is of a much more limited character. It embraces the operations of thought or the rational faculty only in so far as they have to do with the notions and relations of genus, species, individual, etc. It is an arbitrary and what is more an erroneous assumption which claims for Logic the title of *Ars Artium* the [B2.3] Science of which all other intellectual Arts are but applications and which yet in the scientific exposition of that Art recognises no other notions and relations than those to which attention has been directed. In the present essay the term Logic will always be used in its narrower and more technical acceptation. Returning however to the more general philosophy of the intellectual powers to which reference has been made the following principles may I conceive be regarded as of general application viz.:

1st. That Conception has to do with the object of thought only as it falls under some general scientific notion e.g. in Arithmetic under the general notion of Number with its affections in Geometry under that of Figured Space with its affections in Logic under that general notion of Class of which genus and species may be considered affections.—That the analysis of the laws of Conception must consciously or unconsciously precede the analysis of the laws of Judgment and Reasoning.—That none of their laws can be determined *a priori* that is independently of the nature of the concepts involved.—That Judgment consists essentially in a perception of the agreement of concepts but that the nature of this agreement depends upon the nature of the concepts to which its relates; in Geometry for instance it is usually agreement in magnitude, supposed capable of being ultimately tested by superposition, in Arithmetic it is that agreement in respect to number which is termed equality, in Logic the fundamental agreement is as will hereafter be shewn that of identity.—That Reasoning consists in the inferential succession of the propositions by which Judgment is expressed its fundamental principle being that of substitution viz. that two concepts between which an agreement has been established in our premises may in any proposition expressive of the same kind of agreement be substituted the one for the other.—That there are two fundamentally distinct methods in which this principle is applied viz. the method of synthesis and the method of analysis in the former of which we begin with the premises or given propositions and by successive applications of the principle of substitution arrive by a direct process at the conclusion while in the latter we begin by expressing the conclusion in the form of a necessary proposition involving arbitrary elements to be determined and apply the principle of substitution so as to determine these elements in accordance with the premises.—Finally, that the intellectual operations generally conducted by means of their instrument, language, are formal, concerned not with the nature of the individual object of thought but only with the scientific notion under which that object is apprehended and with such notions only as they are of influence and in determining the laws of Thought.

We now proceed to the more detailed statement and application of these principles in relation to Formal Logic.

I. Formal Logic is conversant with things not phenomenologically as objects of perception, nor absolutely as they exist in themselves but only as they fall under the general notions formed by the mental process of abstraction and termed by some of the older writers "logical relations" viz. genus, species, individual etc.

II. These notions are again reducible to that single general notion which may be expressed by the term "Class" and in virtue of which things are capable of being represented by [B 2.4] general names.

III. Logic therefore is conversant with things only as they fall under the general notion of Class and under the several relations or affections of that notion.

The knowledge of external things is in the first instance afforded by perception or at any rate begins with some form of experience. Upon the materials thus furnished the mind operates by its inherent activity. First it forms from those materials the elementary concepts which are expressed by general names and the elementary notions and relations to which those concepts are subject. Secondly it combines these elements together by operations subject to definite laws. It is in the analysis of the second of these processes which begins with concepts and relations *already formed* that the science of Formal Logic according to the views which I have been led to take of it consists. At any rate this analysis forms of itself a very distinct and definite science.

At the same time the analysis of the primary process which explains the genesis of the elements of Logic from the materials which experience supplies is an interesting branch of mental philosophy and to some extent it involves the consideration of the same faculties as are employed in the subsequent operations with which Logic is more especially concerned. Comparison by which we note some particular in which different objects of experience agree. Abstraction by which we fix the attention upon that point of agreement for the exclusion of all other considerations and Generalization by which we conceive of a class of things of which that property shall be the distinguishing mark or attribute seem to be the mental elements involved in the formation of concepts. Sometimes the term Abstraction is used singly and in a larger sense to designate the whole of the process above described.

In every concept thus formed we may distinguish a material and a formal element the former more immediately connected with sense or experience the latter more directly related to the faculties by which we hold converse with the scientific forms of truth. And a further act of Abstraction in the larger sense explained above enables us to attend to this scientific or formal and lay aside the material and sensuous element. Thus if in the first instance the concept "white things" be formed from the perception of "white flowers" "white stones" etc. the concept men from the perception of particular individuals of the race as "Peter" "John" etc., a comparison of these distinct concepts leads us by abstraction to the contemplation of an element in which they agree viz. in that they both involve that notion of Class or Kind which as it has been said is distinctive of the science of Logic.

The proof that the science of Logic is in fact a development of this "notion" will be furnished by the analysis which will follow. But as concerns beforehand the metaphysical possibility of such a notion it may be said that it stands upon the same ground as nearly all other scientific notions e.g. Extension, Substance. It may be and is I doubt not perfectly true that such notions can only be pictured or mentally represented in the concrete individual existence. They may however be contemplated in another way as the ὑπόστασις[5] if I may so speak the subjective unity which constitutes the basis of [B 2.5] formal laws – laws capable of distinct expression and which though manifested only in the special instance are seen nevertheless to possess a necessary and universal truth. When therefore I speak of the general notion of Class with its several affections already enumerated I use that term to designate the common subject of such laws. In reference to that subject, I think the term notion preferable to the term concept because the latter seems to indicate the formation of a mental picture or resemblance. That which is known only by its laws and relations belongs to the understanding rather than to the fancy or imagination; it is a νοητόν[6] rather than an αἰσθητόν.[7]

A doctrine closely resembling that which has been maintained, was held by many of the schoolmen and was expressed in the formula "Logic is conversant with second intentions" (secundæ intentiones). This term was employed to express not as has been erroneously stated* the technical as distinguished from the common use of words but the notions of genus and

---

* The error was first pointed out by Sir W. Hamilton in a review of Whately's Logic. I have taken my own account from Suarez's Metaphysica, Vol. II, p. 516.[8]

species of subject and predicate of antecedent and consequent, etc. Such notions it was affirmed are rational not real, formal not material. They were specially called "second intentions" because formed by a secondary or reflex act of the intellect from the previously formed first intentions given in perception. The term "second intentions" it is observed was not applied to the notions of Genus Species Individual, etc. alone, but to all the abstract notions and relations furnished by the three mental operations of Conception, Judgment and Reasoning.

IV. Whereas in our concepts of things under the general notion of Class, two distinct elements are involved, viz.: 1st.—The representative by which these concepts image forth things. 2nd.—The *noetic* by which such imagery or representation is made in subjection to the general notion of Class; the laws of the intellectual operations with which Logic is concerned are independent of the former and are dependent solely upon the latter or noetic element.

Thus the concepts "men", "stars", "minerals", etc. refer to different classes of material objects but they are all formed by the same intellectual faculty of abstraction from materials furnished by experience and they all embody the same general notion of Class. They might in effect be regarded as different special determinations of that notion. Now the intellectual operations of Conception, Judgment and Reasoning as applied to the above concepts are independent of their special differences [B2.6] and are dependent only upon the general notion which underlies those differences. Thus in forming from the concepts "men" "things white" the concept "white men" we should obey the same laws of the faculty of conception as in forming from the concepts "mountains" and "things burning" the concept "burning mountains". In predicating of men rationality and of stars brightness we should exercise the same faculty of *Judgment*. And the same process of *Reasoning* would be employed in concluding that some general property of the heavenly bodies must belong to Sirius or Aldebaran as in concluding that some general attribute of humanity must be found in the Negro or the Malay.

The principle which has just been illustrated is of fundamental importance in the Science of Logic. But its full value can only be appreciated when through the instrumentality of language Logic is developed into a system of methods and processes which though embodied in signs have their ultimate ground and reason in the laws of the intellectual operations.

V. Our concepts of classes of things the intellectual operations by which such concepts are compounded and the relations under which when connected by propositions they are contemplated can be expressed in a manner adequate to all the purposes of Logic by signs.

We have now arrived at the point of transition from a speculative Logic occupied only with the laws and the relations of abstract thought to a practical Logic based upon a system of signs in whose laws the laws of thought are embodied and by whose combinations their relations are expressed.

Every spoken or written language may indeed be considered as a system of signs adapted to the expression of thought and to the conducting of the intellectual operations in which thought is manifested. And there are certain fundamental principles derived from the constitution of the thinking mind in which all spoken and written languages agree. But in none of the actually developed forms of human speech are these principles found pure and unmixed. They are clogged in their application by the cumbrousness of words. They are distorted by peculiarities of idiom and diversities of construction. For the due and normal development of the Science of Logic it is therefore necessary to lay again the foundations of its sematology where language itself begins – in the processes of thought. To this object we shall proceed in a future section.

As these preliminary observations are intended to be confined to the Philosophy of Logic the determination of the actual laws of the Science being the subject of another part of this paper it may be proper to notice here a question of great importance in the history of logical speculation. What is the real nature and office of signs in the intellectual processes?

The principle which makes the use of signs possible and legitimate is laid down in IV. The validity for instance of the process of *reasoning* depends not at all upon the pictorial element in our concepts whereby the images of sense are reproduced but only upon the formal laws and relations of those concepts. Any mental state or impression which is definite and which can be recalled at pleasure may be substituted for a pictorial concept provided only that it be used in subjection to the same formal laws. The office of external signs is to produce such definite and definitely governed mental state. [B 2.7] Whether these mental states may themselves be considered as signs is merely a question as to the use of words.

If the above is as I conceive it to be a statement of fact, the discussion of the question at issue between the conceptualist and the nominalist will be seen to resolve itself very much into an inquiry about the meaning of words and especially about the *extent* of meaning of the word *sign*. Let us suppose two rival disputants each retaining the language of his sect to seek to interpret that language in accordance with the facts of consciousness. The nominalist must then admit that signs need not necessarily be *external* that they may be mental representations. The conceptualist on the other hand must admit that the concepts which he considers as the immediate objects of the mental regard in Logic need not be actual pictures or resemblances of things. Definite and definitely governed impressions of *any* kind will serve the same purpose. Here then the two doctrines meet. The language of either must be so interpreted as not wholly to exclude the ideas upon which the language of the other is founded. The existence of mental impressions which do not need to bear any resemblance to the outward realities for which they stand, impressions which the nominalist would term signs, which the conceptualist would designate as concepts but which both would admit to be formed by abstraction which both would contemplate as determined to actual use only through the medium of formal laws and relations – this is what I regard as the true basis of the science of Logic.

The objection to the nominalist mode of statement I conceive to be the following. By regarding signs as the only objects of Logic it tends to make us forget that signs even in their most visible and material aspect are still mental in their origin and that their laws belong to them not merely as representative of something external but as formed by an intellectual process of abstraction and deriving from this process their office and constitution. The conceptualist statement is liable to objection on a different ground viz. that it favours a presumption contrary to fact – that we reason only by means of mental pictures or resemblances.

In this as in most other controversies each party seems to confine his gaze either exclusively or by way of preference to a single phase of a many-sided truth. Indeed whenever a theory of phenomena and more especially of mental phenomena is attempted to be condensed into a word that word becoming the symbol of a sect such a result must almost necessarily follow. Even the Realists who gave to the abstract notions of Logic an existence out of the mind and out of language saw but misapplied a great truth viz. that the distinction of genus and species has a real ground in the actual constitution of things which presents to us the spectacle not merely of physical laws maintained but also of organic types preserved unbroken amid cosmical revolutions and through countless ages.

VI. The laws of the signs which express our concepts of things or which express the intellectual operations and relations to which such concepts are subject are independent of the special meaning or content of those concepts and are dependent solely upon the general notion of *class*, which they embody.

This is an immediate consequence of IV. When mental concepts [B 2.8] and mental operations are expressed by signs the laws of those concepts and operations become the laws of the signs themselves. But the laws of the mental procedure are independent of the special meaning or content of the concepts involved and depend only upon the general notion of class under which those concepts fall. Hence therefore the laws of the signs by which that procedure is symbolized are equally general in their application.

Perhaps there is not a single principle in the Science of Logic which has been avowed more generally or in a manner apparently more unequivocal than the above. Aristotle in expressing the terms of propositions by letters set an example of its adoption which nearly all subsequent writers have followed. It is on the ground of this same principle too that modern authorities have described the validity or conclusiveness of arguments regularly expressed as "made evident from the mere form of the expression independently of any regard to the meaning of the words"*. It seems doubtful however whether to its full extent the above principle has been really admitted or even understood by logicians. Its illustrations seem to have been almost exclusively borrowed from the syllogism. Neither am I aware that it has been noted in connection with such statements as have been quoted that the meaning of words is not always wholly independent of the form of the expression in which they occur. Thus the formula "Xs and Ys" does not express an intelligible concept unless the symbols connected by the conjunction *and* be interpreted to signify classes of things wholly distinct. Either symbol indeed taken by way of preference may be considered as arbitrary but when the meaning of one has been fixed that of the other is no longer wholly arbitrary. If by the term "Xs" we agree to mean "mammalia" we cannot interpret the term "s" by "marine animals" because cetacæ which are marine are included in the class of mammalia and the expression "mammalia and marine animals" taken in strictness would be unmeaning. But the condition that "Xs" and "Ys" must represent distinct classes being attended to the interpretation of those terms is in all other respects arbitrary. If we trace to its origin the principle of which the above is not the only kind of example we shall see that the intellectual operations connected with the faculty of Conception do in certain cases impose conditions upon the otherwise arbitrary concepts which are submitted to them and hence it is that the forms of language which is but the outward expression of thought impose conditions of *interpretability* upon the symbols which they connect.

This leads us to the threshold of perhaps the deepest question in the Philosophy of Logic viz. Are we bound when conducting the processes of reasoning by means of language to keep constantly in mind the conditions of interpretability and therefore to employ forms which impose such conditions then only when those conditions are actually satisfied? Is Logic necessarily ostensive in its character? Or is the intellectual procedure in Logic governed solely by a reference to abstract forms and laws? If the latter view be adopted Logic might be described (to use a term which has already been employed) as a noetic not an ostensive science. [B2.9]

I hold, as will be evident from the unfettered statement of VI, the latter view. And the investigation of the laws of the faculties of Conception, Judgment and Reasoning to which I shall shortly proceed will tend I hope to throw upon that view a clearer light than has yet been shed upon it. For I shall while investigating the laws of the intellectual operations determine at the same time the conditions of their interpretability and then shew that the application of the formal laws as a completed system does implicitly and in a very remarkable manner supply the place of that direct consideration of the conditions of interpretability which the ostensive view of the subject would

---

* Whately's *Elements of Logic* Bk. I, Sec. 3.[9]

render necessary. At the same time it is to be observed that whether we adopt the *ostensive* or what I have termed the *noetic* view of the Science of Logic the foundations of that science must equally be laid in an analysis of the laws of the intellectual operations.

VII. The formal development of the Science of Logic requires· 1st. an investigation of the laws of the intellectual operations of Conception as connected with the formation of the terms of propositions 2ndly. an adequate theory of the nature of Judgment by which terms are connected into propositions 3rdly. an analysis in accordance with that theory of the primary laws of Reasoning or of the process which determines the inferential succession of propositions.

It is not necessary to vindicate the above principle which indeed I have stated chiefly with a view to explain the course of the investigations which will follow. The prevalent but very erroneous doctrine which regards all the business and all the interest of Logic as either centered in or converging upon the syllogism has led to a general neglect of the analysis of the faculties of Conception and Judgment. Both the evil and its cause have been made the subject of just comment by Sir W. Hamilton (*Discussions on Philosophy and Literature*, p. 134.)[10]

By an adequate theory of Judgment I mean a theory which while it is based upon a just whether or not it be a complete view of the nature of a proposition is at the same time sufficient for the purposes of logical deduction. It is essentially requisite to such a theory that it should be of uniform application. A theory which would only apply to some particular species of propositions would be inadequate. The primary laws of reasoning or of the inferential succession of propositions must in expression at least depend upon the view which is adopted with respect to the nature of a proposition.

VIII. The form which the developed Science of Logic must assume cannot be determined *a priori* but must depend upon the nature of the laws established by a direct analysis of the intellectual operations.

When it is said that such an analysis of the intellectual operations as I have described is necessary to the formal development of the Science of Logic no principle is asserted which is not equally applicable to every other department of scientific inquiry. The form of a science can never be determined *a priori*. A special and careful analysis of that which constitutes its matter is an essential prerequisite in the first place of all valid reasoning concerning that matter in the second place of all legitimate speculation concerning the nature form and method of the science of which it forms the [B2.10] subject. It would betray a rash self-confidence if we should endeavour to prescribe the forms and conditions of the Science of Dynamics without a previous enquiry into the laws of motion, or of the Science of Morals without an analysis of the nature and the grounds of moral obligation.

Analysis of the Operations of Conception as Exercised within the Sphere of Formal Logic

As Logic is concerned with *things* under the general notion of Class, we must in the analysis of the faculty of Conception resolve all elementary concepts into concepts of things. The proposition "Stones are heavy" must be contemplated under the form "Stones are heavy things" for thus and thus only it exhibits a relation between *things* viewed under the general notion of *Class*. Here it is proposed to determine in what manner and in subjection to what laws and conditions new concepts are formed from elementary concepts previously given. How these elementary concepts are formed has been explained in ( ).[11]

According to a recognised division the concept of a class of things may be considered 1st. with reference to its extension as a whole made up of parts 2ndly. with reference to its intension as formed by the union or combination of qualities common to all the individuals which it comprehends. The concept "minerals" as a class of things including gold, silver, iron, aluminium etc. is viewed in extension; the concept "minerals" as involving the qualities of ductility, fusibility, a peculiar lustre etc. qualities common to all the individuals of the class "mineral", is viewed in intension.

Now the elementary operations of Conception are four in number two of them founded on the extension two on the intension of a class. And of these four operations each pair consists of operations mutually inverse to each other. All this will appear from the definitions of the operations to which we next proceed.

Operations founded on Extension

1st. Addition by which from the concepts of parts we form the concept of the whole which they constitute as when from the concept "men" and the concept "woman" we form the concept "men and women". The sign of this operation in language is the conjunction "and".

2nd. Subtraction or the inverse of addition viz. that operation by which from the concepts of a whole and the concept of one of its parts we form the concept of the other part, as when from the concept "stars" and the concept "planets" we form the concept "stars which are not planets" or "stars except planets". The sign of this operation in language is the conjunction "except".

Operations founded on Intension

1st. Composition by which from the concepts of two classes of things we form the concept of that class which consists of all the individuals which are common to the two classes as when from the concept "men" and the concept "white things" we form the concept "white men". [B 2.11]

2nd. Abstraction defined as the inverse of composition viz., that operation by which from two given concepts one of which enters into the other by way of composition the other component concept *in its utmost generality* may be determined. The term abstraction in its usual acceptation (and it is thus that I have employed it in the introductory portion of this paper),[12] denotes a particular and special form of the operation which has just been defined. For instance when from the concept of "white flowers" and that of "flowers" we form the concept "white things" we indeed arrive at the concept of a class of things from which if we mentally select those which are "flowers" we arrive at the original compound concept "white flowers". But this concept "white things" is not the only one from which by composition with that of flowers the concept "white flowers" may be formed. To the concept "white things" we might add the concept of any class of things possessing neither whiteness nor the floral character or of any indefinite portion of such a class of things and still the resulting concept would be one which by composition with [that] of "flowers" would generate the concept "white flowers". Thus if from that class of things which consists of "white things together with red leaves" we mentally select those individuals which answer to the description "flowers" we arrive at the concept "white flowers". Or if from the concept "white things" we subtract the concept of any class of things which while white are not flowers or of any indefinite portion of such a class of things we still arrive at a concept which by composition with "flowers" generates the concept "white flowers". When therefore abstraction is defined as the inverse of composition its office must be regarded as the determination not merely of some particular and definite concept answering a given requirement but of some general form of thought expressible in language and including all possible concepts however indefinite in extent by which the same requirement is answered.

It is obvious from what has been said that the operation of Abstraction differs in some important respects from the operations of Addition, Subtraction and Composition. When by mental addition we form the concept of a whole out of the separate concept of its parts, or by the inverse operation of subtraction that of a part from the previously given concept of the whole and of the remaining part no individual can be classed under the final concept which was not present to the mind in the concepts originally given. There is no individual under the class "men and women" which is not comprised under one or other of the separate classes "men", "women", neither is there any individual in the concept "men" as a part which was not comprised under the initial concept "men and women" as a whole from which the part is derived by mental subtraction. So the operation of composition introduces as is obvious no individuals not comprised in the initial concept to which it is applied.

But the operation of abstraction brings before us individuals not previously contained within the sphere of Conception. When from the concept of "volcanoes" we form that of "mountains" by abstraction of the quality "burning" we are led to the contemplation of things not contained in the initial concept. The class "mountains" includes individuals not comprised in the class "volcanoes". In this respect as well as in the indefinite extent already noticed of the concepts to which in its most general acceptation it leads the operation of Abstraction differs from the three others with which it is connected.

It is to be noted that the terms addition, subtraction, etc. [B2.12] are applied to mental operations only in a derived and metaphysical sense. Thus it is nowhere meant to be affirmed that we can add two concepts together in the same way as we can collect into one whole two physical parts. But the term addition when employed to represent an operation performed upon the concepts "trees", "herbs", means only that operation whatever it may be by which from these concepts we form the concept "trees and herbs". We can in no other way describe mental operations than by terms material in their origin and here and throughout this paper the primary meaning of the operations described is to be sought for in their physical analogue. Should it be asked—Why then the introduction of mental operations and relations is necessary it is replied that they possess a truth and application independently of the conditions under which their physical or objective realization is possible. [B2.12a]

Laws of Conception

Addition

LAW. *In forming the concept of a whole from the concepts of its parts by addition whether 1st immediately or 2ndly mediately by connecting any of those parts into partial wholes and then connecting those partial wholes into the final whole the order in which the several mental acts are performed is indifferent.*

Thus in forming the concept "trees and herbs" we can either begin with the concept "trees" as the primary object of thought and then by mental addition of the concept "herbs" form the concept of the whole "trees and herbs" or we can begin with the concept of "herbs" and then by mental addition of the concept of "trees" form the equivalent whole "herbs and trees". I say *equivalent whole*, for while the concepts "trees and herbs" "herbs and trees" differ in the order of the intellectual operations by which they are formed and in the order in which the partial concepts which they involve are presented to the mind they agree in their material signification and in their logical value. By this latter expression I mean that the one form of the concept may in any process of reasoning be substituted for the other.

Again if we had to connect together the partial concepts "trees" "grasses" "mosses" "ferns" into a single total concept by addition we might either do this immediately by adding to the concept of "trees" that of "grasses" to the resulting concept that of "ferns" and to the result thus obtained that of "mosses" or we might form first the concepts of the partial wholes "trees and grasses" "mosses and ferns" and connect these by addition. Or any other grouping or arrangement may be adopted which takes into account all the partial concepts and connects no parts which are not distinct.

CONDITION. *In order that two or more concepts may be formed into the concept of a whole by the operation of addition the classes of things which these concepts represent must be wholly distinct.*

This condition has been already exemplified and it is only stated for the sake of formal completeness.

Subtraction

Subtraction is the inverse of addition. For as we pass from the concept of "trees" to that of "trees and herbs" by mental addition of the concept "herbs" so by mental subtraction of the same concept "herbs" we revert from the concept of "trees and herbs" to that of "trees".

It would be easy but it is unnecessary to investigate the laws of subtraction. For as the process of subtraction is the inverse of addition and as it is fully defined by that relation its laws are necessarily founded in the law of addition and this has been determined already. We might indeed have in the most general case to consider a series of mixed operations some belonging to addition others to subtraction and with reference to any such series the following law may be established viz.: [B 2.13]

LAW. *Whenever a concept is formed from other elementary concepts by the operation of the mental faculties of addition and subtraction the order of possible succession of the mental acts is indifferent.*

Thus the concepts expressed in the two following forms are equivalent though arising from different but possible series of mental acts.

"Trees and herbs except pine trees"

"Trees, except pine trees, and herbs".

But this law may as I shall hereafter shew be deduced as a consequence of the Law of Addition already established and of the definition of subtraction as the inverse of addition and it is therefore not to be ranked among those laws which in the scientific exposition of logic must be regarded as primary. At the same time it is to be remembered that the principle of classification which makes addition a direct and subtraction an inverse operation is *in some degree* arbitrary. That the operations are mutually inverse is clear, that we should avail ourselves of this relation for diminishing the number of the laws of thought which are to be considered primary is a scientific necessity, and few will doubt that if one or the other of the mutually inverse operations is to be regarded as direct the preference should be given to addition. Still it is not to be forgotten that the operation of subtraction is intelligible in itself and that its laws may be independently determined.

CONDITION. *In order that any concept may be mentally subtracted from another concept it is necessary that there should be comprehended under the former concept no individuals which are not comprehended under the latter.*

Thus the expression "Trees except pines" is interpretable in thought because and only because pines are understood to be a species of trees.

Composition

LAW 1st. *When concepts are combined by composition whether immediately or mediately the order in which the several mental acts of composition are performed is indifferent.*

Thus in forming the concept "white flowers" from the elementary concepts "white things" "flowers" we can either begin with the concept "flowers", then mentally fix the regard upon those of the class which also come under the general concept "white things" or we can begin with the concept "white things" and then mentally select those individuals of the class which fall also under the concept "flowers". The concepts to which we are led differ only formally. In material reference and in logical value they are equal. Of the law of Conception above stated abundant illustrations are found in language. The indifference of the order in which in any language two adjectives absolute in their meaning and applied to the same subject succeed each other affords a familiar example. To say "white scented flowers" is equivalent to saying "scented white flowers". Of the permitted inversions of structure in the Greek and Roman languages a large portion (perhaps all that are not idiomatic) depend either upon the above or upon some cognate principle.

It is further evident that in Composition it is indifferent whether elementary concepts are combined immediately or mediately. Thus in combining the concepts expressed by the terms [B2.14] "white" "scented" "fading" "flowers" it would be permitted to form first the compound concept "white scented things" secondly that of "fading flowers" and then by a final act of composition combine together these compound concepts. This is that method of composition which I have termed *mediate*.

LAW 2ND. *When the two elementary concepts which are compounded are the same so that one might be regarded as a reproduction of the other, the concept formed by their composition is also the same as either of the original concepts.*

For in compounding two concepts we form the concept of the class of things which is composed of the individuals common to the classes represented by the component concepts. When these component concepts therefore agree in representing the same class the individuals of that class stand in the place of the individuals which are common to the two classes when the concepts are different. Wherefore the concept formed by composition is the concept of the same individuals and in no respect differs from the concepts from which it is formed.

This law is exemplified in language. To say "white white flowers" though pleonastic is equivalent to saying "white flowers". The office of the adjective when applied to the expression of any class of things is not simply *attributive*, it does not present to us the concept of the same class with an attribute of quality attached to every individual of that class. To speak of "good men" is not to speak of all men as good. The office of the adjective in common discourse involves two elements, mental selection as well as mental attribution. It directs us to fix the attention on those individuals of a class and on those only to which a certain attribute is applicable. Thus the adjective "white" as applied to the class "flowers" while it raises the concept of whiteness directs us to fix the attention on those members of the class "flowers" which possess that quality and the reduplicate form "white white" applied to the same class, does but repeat the direction. (Laws of Thought, p. 32.)*

The operation of composition is not subject to any condition analogous to those which accompany the operations of addition and subtraction. If two classes have no individuals in common, the composition of their concepts produces the concept "no individuals" or "nothing". And this is in itself an intelligible concept.

Abstraction

As abstraction is the inverse of composition and is fully [B 2.15] defined by that relation, its laws do not as is evident from what has been said with

---

* On the passage referred to in the Laws of Thought the following note communicated to me by Dr. Latham is interesting. Referring to the law exemplified in such expressions as "good good" he observes "The practice of language gives us better instances in the persons of verbs in those languages where there is besides the sign of the person as an inflection the personal pronoun itself as an adjunct.—Ego su-m, I a-m."[13]

reference to the inverse operation of subtraction require to be separately determined. They do not hold in the scientific exposition of Logic the rank of primary laws. The entire theory of the operation of abstraction will be seen to flow from the laws of the direct operations which have already to some extent been considered and from the general principles of reasoning to which attention will shortly be directed.

CONDITION. *In order that one concept may be abstracted from another it is necessary that the class represented by the former concept should be entirely comprehended in the class represented by the latter.*

Thus an ostensive application of the operation of abstraction is involved in the abstracting of the concept mountains from the concept "volcanoes" defined as "burning mountains" the result (or rather a special result) of this operation being the concept "things which burn". Now all the individuals comprised in the class "volcanoes" are comprised in the class "mountains". And supposing even that the definition of "volcano" were not given the *possibility* of abstracting from the concept "volcanoes" the concept "mountains" rests upon the supposition that the concept "volcanoes" is comprised under that of "mountains" for the object of that operation is the discovery of the concept of a class of things from which if we select those individuals which are mountains we shall obtain the concept volcanoes. And this implies that under the concept mountains that of volcanoes must be comprehended.

Law of mixed operations of Addition and Composition

From the possibility of conducting a mixed series of operations arises an important law which is entitled to be ranked as primary. It may be thus stated.

LAW. *When any concept formed by addition is employed in the way of composition it may either be employed as a whole or the parts which it involves may be employed separately and the resulting concepts connected by addition.*

Thus the expression "aromatic trees and herbs" is equivalent to the expression "aromatic trees and aromatic herbs". In the former expression the concept "aromatic" is applied by way of composition to the whole "trees and herbs" formed by addition of the elementary concepts "trees" "herbs". In the latter expression the concept "aromatic" is compounded separately first with the concept "trees" thus forming the concept "aromatic trees" secondly with the concept "herbs" thus forming the concept "aromatic herbs" finally these results are connected together by the operation of addition. It is evident that the law is general and involves no limitation as to the number of the elementary concepts connected together.

Analysis of Judgment as Exercised
within the Sphere of Formal Logic

Judgment has been defined as that power of the mind by which propositions are formed. I do not purpose to enter here into an examination of the different species of propositions but shall confine myself to a single question viz. what is [B 2.16] the nature of a proposition? Nor shall I discuss this question any further than is necessary to determine what definition of a proposition is singly sufficient as a basis for the theory of reasoning. At present also it will be enough to consider that family of propositions which is termed by logicians *categorical*.

Two distinct views of the nature of a proposition have been maintained. By some a proposition is regarded as a form of predication by others as an expression of identity.

The former of these views seems most directly applicable to those propositions of ordinary discourse in which the term called the predicate is to use the language of the schools *particular* more especially if it be expressed by an adjective. When we say "snow-flakes are white" it seems most natural to regard this expression as a form of predication by which we attribute to a certain subject "snow" a certain quality perceivable directly by sensation "whiteness". When agreeably to what has been said in the introductory remarks on Conception we contemplate the proposition under the form "Snow-flakes are white things" the notion of predication though not quite so prominent as before is still perhaps more so than the notion of identity.

On the other hand the notion of identity seems most obviously to be presented when the terms connected by the copula *is* or *are* are both universal whether this universality is due to both terms being singular as in the proposition "Louis Napoleon is Emperor of the French" or to both terms having the definite article prefixed as in the proposition "the contented are the truly rich" or lastly to the proposition itself being understood as a definition e.g. "Triangles are plane figures bounded by three straight lines."

I have said that the above are the most direct and obvious views of the nature of the proposition in the different cases which have been considered. It is however possible that besides that view which is in each case the most obvious one the other which is the less so may be really involved also. A proposition may possibly have two aspects in the one of which it may appear as a form of predication, in the other as an expression of identity. This question I purpose next to consider.

It has been observed by Professor De Morgan that every proposition which appears in the form of an identity is resolvable into two others in which predication is involved.[14] Thus the proposition "The truly rich are the contented" while in its plain and obvious sense it expresses the identity of the two classes "the truly rich" and "the contented" is resolvable into the two predications. "The truly rich are contented" and "The contented are truly rich."

On the other hand every proposition which is expressed in the form of a predication may be also expressed as an identity. The principle of the "quantification of the predicate" which forms the basis of Sir W. Hamilton's logical scheme serves to effect this reduction. The proposition "Lilies are white" may be adequately replaced by a proposition affirming the identity of the class "lilies" not indeed with the class "white things" but with some indefinite and unknown portion of that class. The theory of this reduction may be well explained in the language of the old scholastic logic. Thus [B 2.17] the proposition under consideration may be understood as implying that lilies are a species included in the genus "white things". But "species" is formed by the composition of *genus* and *differentia*. There exists some characteristic property or collection of properties *differentia* by which lilies are distinguished from other white things and which if known would in conjunction with the common property of whiteness constitute the definition of lilies. When the differentia is as we have supposed it to be in the present case unknown its place may be occupied in language by the indefinite pronoun *some*. Thus we may say "Lilies are some white things" the copula *are* being here employed not in the sense of predication but of identity.

It appears then that while some propositions *appear* as forms of "predication" others as expressions of "identity" either of these views of the nature of a proposition may be adopted as the basis of a logical system. The view which regards a proposition as an expression of identity is however the more direct in its application inasmuch as any proposition which appears in the form of a predication becomes by merely expressing the understood quantification of its predicate an identity whereas identities which connect universal terms cannot be represented by single forms of predication but only by a combination of such forms. There exist also considerations yet deeper than these which leave no doubt what view of the nature of a proposition ought to be adopted here. Logic it has been said is concerned with things not as they exist in themselves but only as they fall under the general notion of Class. The laws of Conception which might in fact be termed the laws of the formal development of the notion of Class have been investigated in a previous section of this paper. Now these laws if exemplified at all in the form of propositions are without exception exemplified in the form of identical propositions. When in accordance with the primary laws of addition we affirm that the concept "trees and herbs" is equivalent to the concept "herbs and trees" we mean that those concepts though differently formed represent and *equally* represent the same class of external things. If the sign = be employed to express identity we shall have the identical proposition

Herbs and trees = Trees and herbs.

And in such identical propositions as the above may all the laws of the faculty of Conception be exhibited.

Upon the ground of analogy far more deeply founded than analogies usually are we may therefore conclude that in the scientific development of logic the proposition is to be regarded as an affirmation of identity not as a mere form of predication. And this view of the Subject will be confirmed by the results of the analysis of the operation of Reasoning to which object we shall next proceed. [B 2.18]

Analysis of Reasoning as Exercised within the Sphere of Formal Logic

The office of Reason is to determine the inferential succession of propositions. Moreover a proposition is in Formal Logic an affirmation of identity; it asserts that it is to the same object—to the same individual *things*—that two concepts in thought or two terms (properly limited) in language, refer. Logical Reasoning therefore is specially concerned with the notion of identity and it might with propriety be said to consist in the operations which arise from the development of that notion. But as the identity with which logical reasoning is concerned is the identity of the *things*, which as represented by the terms of the proposition have become the subjects of Conception under the general notion of *Class* and in subjection to the formal laws and relations therewith connected we are ultimately conducted to the following definition viz.: Logical reasoning is a process of *formal* inference applied to propositions considered as affirmations of the *identity* of the things which as conceived by us under the notion of [C 42.1] Class are expressed in the connected terms of propositions. We shall in speaking of the connexion of the terms of propositions or of the relation between the concepts which those terms express describe that connexion or relation as one of *equality*—equality of two terms or concepts corresponding to identity in the objects which they represent.

The elements in the above description to which it is most important to attend and from which the laws of Reasoning as a process are derived, are the following viz.:

1st. Reasoning (in Logic) is a process founded upon the notion of identity.

2nd. Reasoning is a formal process.

Now the fundamental law of Reasoning as developed from the notion of identity is the following:

"Two concepts which in the sense above explained are equal to a third concept are equal to each other."

And the rule or principle regulative of the process of Reasoning founded upon this law is the following viz.:

Terms or concepts which are equal may be substituted the one for the other in any process of logical inference.

As an example of the application of this prin- [C 42.2] ciple, which may be termed the principle of substitution, suppose the two following premises given

Similar triangles =
triangles which have their respective angles equal.

Similar triangles =
triangles which have their respective sides proportional.

Then it follows from the principle of substitution that

Triangles which have their respective angles equal =
Triangles which have their respective sides proportional.

The order of the process followed in the above example consists in substituting for the concept "similar triangles" in the second premiss the equal concept "triangles which have their respective angles equal" derived from the first. It may be observed though the remark does not affect the character of the reasoning that neither of the premises is a mere nominal definition. There exists a distinct definition of "similar figures", which involves consideration neither of angles nor of sides. Neither is the reasoning, though it relates to a geometrical subject geometrical reasoning.

There are three applications of the direct law of substitution which admit of distinct exhibition in the form [C42.3] of subsidiary or derived laws of reasoning. They are as follows.

LAW A. *If to equal concepts the same or equal concepts be added, the resulting concepts are equal.*

LAW B. *If from equal concepts the same or equal concepts be subtracted, the remaining concepts are equal.*

LAW C. *If equal concepts be compounded with the same or equal concepts, the resulting concepts are equal.*

In the "Laws of Thought" these laws appear as primary. I believe that if carefully examined it will be seen that the ground of the intellectual assent to their truth is founded in the perception that they are direct applications of the law of substitution a perception which becomes the stronger from the comparison of the different laws with each other.

Thus suppose it given that

"Philosophers = lovers of wisdom"

An example of Law C. would be furnished by the inference that

"Philosophers who are poets = lovers of wisdom who are poets", and it is plain at least that this *may* be obtained by the substitution in its first member or subject of the term "lovers of wisdom" for the term "philosophers" the equality of [C42.4] those concepts being guaranteed by the premiss.

Again, if we had the premises

"Practical men =
men who value the ideal only for the sake of the real",

"Speculative men =
men who value the real only for the sake of the ideal",

we should have as an illustration of Law A:

"Practical men and speculative men =
men who value the ideal only for the sake of the real and men who value the real only for the sake of the ideal."

It is manifest that this is equally with the last illustration an example of direct substitution and if we compare the two examples together it will appear that this is the only *element* in which they agree. The operations by which the first members are formed are quite distinct. The one is formed by Composition the other by Addition. Now the nature of the operation no further affects the reasoning than as such operation furnishes a ground for the application of the principle of substitution. The syllogism as it perhaps most readily presents itself to the common sense of mankind seems to rest on a similar basis. Take for example the syllogism [C42.5]

"Men are mortal beings
Kings are men
Therefore kings are mortal beings"

Reducing the premises to the form of identities

"Men = some mortal beings
Kings = some men"

Therefore by substitution

"Kings = some some mortal beings";

that is to say, "Kings = some taken out of some mortal beings", and this is perceived to be the equivalent to saying "Kings = some mortal beings" or "Kings are mortal".

If we carefully examine the above process of inference we shall see that it involves two distinct steps viz.: the act of substitution leading to the formal conclusion "Kings = some out of some mortal beings" and the further process dependent upon Conception by which this conclusion is brought to the form "Kings are mortal". It is apparent even from this example that the principle of substitution is not of itself sufficient for the conducting of an argument. It must be employed in combination with other intellectual acts dependent upon the faculties of Judgment and Conception and it is in such combination that it is [C 42.6] actually employed in the ordinary reasonings of mankind. To these conditions of its application it may be added that it of necessity fails us when as in all the more difficult forms of logical reasoning we have to call in the aid of Abstraction. For the operation of Abstraction is as it has been seen purely inverse and therefore dependent in its character and essentially involves the employment of those faculties of the mind by which necessary propositions are formed. And hence while we are able by the principle of substitution alone to combine the premises together through operations of Addition Subtraction and Composition we are compelled to have recourse to those canons of abstract thought which enable us to add to the premises whose truth is only assumed other propositions whose truth is not assumed but necessary in other words we must adopt the analytical and not the synthetical method before we can apply the same principle of substitution in connexion with the operation of Abstraction. The application of the analytical method is however so dependent upon language and its exposition is so much facilitated by the employment of a proper system of notation that it becomes if not necessary at least highly [C 42.7] important to introduce such a system and avail ourselves of its aid in expression before proceeding further into the analysis of Logical Reasoning.

It is important also to present the laws of the intellectual operations not as isolated truths but as constituent parts of a system governed by pervading relations; and to this object the employment of an adequate system of notation is equally requisite.

On Systems of Notation

The ground of every system of notation employed in reasoning is the formal character of reasoning itself. If the process of inference is independent of the particular meaning of the concepts involved and depends only upon the general notion which those concepts manifest and thereupon only as it furnishes the basis of intellectual operations and of formal laws it is at once suggested to us that we express concepts not as in ordinary language by words the special meaning of which may through association of ideas interfere with our perception and application of the purely formal laws to which they are subject but by symbols in the employment of which the formal law and not the special meaning is present to the mind. It is not to be forgotten [C 42.8] by those who object to symbolical representation that the founder of Logic set the example of their use.

The excellence of a notation consists in this that it expresses directly by its elementary symbols the elementary concepts operations and relations of the system of thought to which it belongs. In the system of Algebra there exists but one kind of elementary concept that of number but four operations by which the concept of number can be modified viz. Addition Subtraction Multiplication and Division and (in the most important applications of the science) but one relation — that of equality by which propositions connecting these concepts are expressed. These elements of Conception operation and relation are directly expressed by symbols viz. concepts of number if general by letters if particular by the Arabic numerals, the operations of Addition Subtraction Multiplication and Division by the respective signs + − × ÷ and the relation of equality by the sign =. The perfection of the language of Algebra is due to the circumstance that it has been found possible thus to determine beyond all question or contradiction the ultimate elements of thought in the system to which that language is applied. For expressing these elements by signs it follows that all their combinations [C 42.9] possible in thought will be expressible by combinations of signs in subjection to formal laws which represent the laws of combination of the original elements. We are acquainted with the origin of the peculiar signs of Algebra and know how they were evolved from common language. The form of the sign + for instance originated in the *et* and the sign minus in a mark of contraction employed in expressing the word *minus* in mediæval latin manuscripts. But the ideas which these signs express had been formed previously and their truly elementary character in reference to the system of number perceived.

The formal laws of Algebra usually relate to the permitted order of succession of operations. We know for instance that if the number 7 be multiplied by 8 the result is the same as if the number 8 be multiplied by 7 viz. 56; and we know that this is not due to any peculiarity in the numbers 7 and 8 but only to the fact that they are numbers. If x and y represent any two numbers whatever and if xy represent the product obtained by multiplying the number x by the number y we shall have as a general law

$$xy = yx \qquad\qquad [C 42.10]$$

In like manner we have as a formal law of Addition

$$x+y = y+x$$

and as a formal law of operations in which Addition and Multiplication are mixed

$$x(y+z) = xy+xz$$

the latter expressing that if we add any two numbers y and z and multiply the sum by a third number x the result is the same as if we first multiplied y and z separately by x and then added the products.

To these which are properly speaking the fundamental laws of Conception in Arithmetic we must add that great law of substitution the basis of algebraical reasoning in virtue of which any combination of symbols may be substituted for any other combination of symbols known to represent the same value.

Now the elements of Logic are not less definite than those of Algebra. We have but one kind of elementary concept viz. that of class — but four elementary operations by which concepts of class can be modified viz. Addition Subtraction Composition and Abstraction but [C42.11] one fundamental relation expressed by propositions viz. the relation of identity. If we express these elements by symbols of conception operation and relation corresponding to the symbols of Algebra an equally definite system if not the same system of formal laws will govern their employment. Those laws have been already investigated in previous sections of this paper and it only remains to give to them their formal statement and consider them as the systematic basis of a developed scheme of Logical Science. To this statement we shall now proceed.

Symbolical expression of the formal laws of Logic

It is by no means essential to adopt in the expression of the forms of Logic the symbols employed in Algebra. It is however a matter of convenience to do so. For although the ideas embodied in the symbols of Algebra are for the most part different from those which the same symbols would embody if employed for the expression of propositions in Logic yet does there exist between the two sciences such a fundamental relation (however it may be explained) that the formal laws of the symbols are with one exception the same in the two systems. And there exists a special Algebra (i.e. [C42.12] a science which has to do with ideas of number but not with *all* ideas of number) of which the formal laws and therefore the processes are in all respects identical with those of Logic.

I have said that the ideas of Algebra are not for the most part the same as those of Logic but some of them (though this is not at all important to the general argument which rests only on the comparison of formal laws) are apparently the same. Thus the ideas of Addition and Subtraction in the two sciences seem to agree essentially as well as formally. And this suggests at once the employment of the same symbols for their expression.

Let letters be used to denote classes of things as subjects of conception and let the four elementary logical operations of Addition Subtraction Composition and Abstraction be expressed by the same signs as the respective arithmetical operations of Addition Subtraction Multiplication and Division. Thus—Let $x+y$ denote the class formed by adding the members of the class x to those of the class y supposed distinct. E.g. if x represent "Trees" and y "Herbs" let $x+y$ represent "Trees [C42.13] and herbs".—Let $x-y$ denote the class formed by subtracting from the class x the class y supposed to be wholly contained therein. E.g. if x represent "Men" and y

"Negroes" let $x-y$ represent "men who are not negroes".—Let $x \times y$ or xy denote the class whose members are common to the classes x and y. Thus if x denote "Flowers" and y "White things" let xy denote "White flowers".—Let $x \div y$ or $\frac{x}{y}$ represent that class of things from which if we select those which belong to the class y we shall obtain the class x.—Let the symbols 1 and 0 be respectively used to denote the Universe and Nothing. The ground of this selection is that the symbols 0 and 1 are subjected to the same formal laws when thus interpreted in Logic as when employed in Arithmetic. Hence the expression $1-x$ will denote that entire class of things which remains after taking away from the universe the class denoted by x i.e. it will denote the class of things which are not members of the class x.—Let the placing of any expression within brackets or under a vinculum denote that it is to be treated according to the same laws as if it were a single [C 42.14] letter representing a class. Thus let $y(1-x)$ or $y\overline{1-x}$ denote the class of things which consists of all individuals that are found in the class y but not in the class x.—Let the sign = interposed between the expressions of two concepts denote that these concepts are equal in the sense explained in ( ) viz. that the classes which they image forth consists of the same collection of individuals. Thus if x represent "Men", y "Rational beings" and z "Animals" the equation

$$x = yz$$

will express the proposition

"Men and rational animals are identical".

It is possible to express by the above notation any categorical proposition in the form of an equation. For to do this it is only necessary to represent the terms of the proposition regarded as subjects of Conception by symbols introducing if needful an indefinite class symbol and then to connect the expressions thus formed by the sign of equality.

Thus if it were required to express the proposition "Trees are plants" we should have representing "Trees" by x, "Plants" by y, and the indefinite aggregate of those qualities which distinguish trees from other plants by v, the equation [C 42.15]

$$x = vy$$

If it were required to express the proposition "Stars are celestial bodies which either are self-luminous and do not shine by reflected light or shine by reflected light and are not self-luminous" then representing

      stars by s
      celestial bodies by c
      self-luminous bodies by a
      bodies shining with reflected light by r

we should have the equation

$$s = c\,(a\overline{1-r} + r\overline{1-a})$$

or
$$s = vc\,(a\overline{1-r} + r\overline{1-a})$$

according as our proposition was intended as a definition or as a mere description of stars. In the former case it would be implied that the class of things called stars is identical with the class of things whose properties are described in the second member. In the latter it would only be implied that the class of stars is included in that class the nature of the further distinction by which stars are recognised being left undetermined.

When the terms of propositions and the relations between those terms by which propositions are constituted are [C42.16] expressed by symbols the laws of the faculty of Conception by which these terms are formed become the laws of the symbols employed and the laws of Reasoning are interpreted into axioms.

The adequacy of this system of notation as a means of expression having been established it remains to exhibit the primary formal laws of its symbols. For this purpose it is only necessary to translate into its language the laws of Conception already determined in ( ).

The formal law of Addition will be expressed by the equation

$$x+y = y+x \qquad \text{(I)}$$

The formal laws of composition are the following

1st $\qquad\qquad xy = yx \qquad\qquad$ (II)

2nd $\qquad\qquad xx = x \qquad\qquad$ (III)

The formal law of mixed operations of Addition and Composition is

$$x(y+z) = xy + xz \qquad \text{(IV)}$$

The above are the only primary laws of Conception. For the laws of inverse are derivable from those of direct operations. The above therefore determine the character of the system so far as the formal laws of Concep-[C42.17] tion are concerned.

If we compare the laws with those of Algebra as formally expressed in $\int\int$ [15] we see that I II and IV viz. the law of Addition the first law of Composition and the law of mixed operations of Addition and Composition agree severally with the law of Addition, the law of Multiplication, and the law of mixed operations of Addition and Multiplication in Algebra. Hence the general rules of Algebra e.g. the law of signs in multiplication etc. remain formally true here. The second law of Composition has however no analogue in Algebra as the science of numbers in general. The formal difference between Logic and Algebra consists then in this that the Concepts with which the former science has to do are subject to a peculiar law symbolically expressed by the equation

$$xx = x \qquad\qquad (\ )$$

Now there exist two numbers viz. 0 and 1 which besides satisfying the general laws of Algebra satisfy also the above special formal law of Logic. If then we construct an Algebra in which the only particular symbols of number shall be 0 and 1 and in which every general symbol as x, y etc. shall be understood to admit only of the above special determination (i.e. it being given that x is a [C42.18] literal symbol belonging to the dual Algebra it shall thence be understood that x means either 0 or 1 but it is undetermined which) the formal laws of such an Algebra will be identical with those of Logic when expressed by symbols. Thus we shall have as particular manifestations (\ )

$$1 \times 1 = 1$$

$$0 \times 0 = 0$$

no other numerical symbols than 0 and 1 obeying the same law. And hence the processes of the dual Algebra in question (dual because recognizing no other numbers than 0 and 1) will be formally identical with the process of Logic expressed by symbols.

And this agreement extends not only to the formal laws of operation in the systems compared but also to the formal conditions of interpretability of expressions formed by those operations. That three out of the four operations of Conception in the logical system viz. Addition Subtraction and Abstraction are subject to such conditions has already been seen. I shall therefore shew that these conditions are capable of formal expression and that they are when thus expressed common to the systems of Logic and dual Algebra. The ground of this agreement will further be shewn [C42.19] to consist in the community of the formal laws of operation and more especially of the distinguishing law (III).[16]

In fact representing any combination of symbols by V, whether that combination express a logical concept or a concept in dual Algebra, i.e. one of the numbers 0 and 1, it must by what has preceded equally satisfy the law

$$VV = V \qquad\qquad (1)[17]$$

First then suppose V to represent the combination $x+y$ wherein x and y obey the same law so that

$$xx = x$$

$$yy = y$$

Now writing $x+y$ for V (1) becomes

$$(x+y)(x+y) = x+y$$

Performing the operation indicated in the first member and replacing in the result xx by x and yy by y we have

$$x+xy+yx+y = x+y$$

whence $\qquad\qquad xy+yx = 0$

an equation which in either system can only be satisfied by supposing

$$xy = 0 \qquad\qquad\qquad (2)$$

since in Logic the only class, and in dual Algebra the only number, which being added to itself produces Nothing is [C42.20] Nothing. Hence the equation $xy = 0$ is the condition of interpretability of the operation of adding y to x in either system and it is a condition formally expressed.

Logically interpreted this condition demands the non-existence of the class xy. Now to demand the non-existence of the class xy is the same as to demand that the classes x and y shall be wholly distinct. And such is the condition of possibility of the actual operation of Addition already determined in ( ).

Interpreted in the system of dual Algebra the condition $xy = 0$ demands that the values of x and y should be so chosen that their product should vanish. And this restricts the actual selection to the following pairs of values viz.:

1st	$x = 1$	$y = 0$
2nd	$x = 0$	$y = 1$
3rd	$x = 0$	$y = 0$

and excludes the combination $x = 1$ $y = 1$. Now if we substitute these combinations in the expression $x+y$ we find the three first give to that expression the respective values 1, 1, 0, values belonging to the dual system while the last combination gives to the expression the value 2 which is not included in that system. The sufficiency of the [C42.21] formal condition $xy = 0$ for the interpretability of the expression $x+y$ is therefore established in both systems. And this condition is derived from the fundamental law of Conception ( ).

If we apply the same analysis to the expression $x-y$ formed by Subtraction we shall arrive at the formal condition of interpretability

$$y(1-x) = 0$$

Logically interpreted this condition demands that there should exist no individuals in the class y which are not found in the class x. Such is the condition already assigned in ( ). Interpreted in the dual Algebra it would permit the combinations

$x = 1$	$y = 1$
$x = 1$	$y = 0$
$x = 0$	$y = 0$

and exclude the combination

$$x = 0 \qquad y = 1$$

Now the first three of the above combinations substituted in the expression $x-y$ give to it the respective values 0, 1, 0, which are included in the dual system. The last combination gives the value -1 which is not included in that sys- [C 42.22] tem.

Applying the same analysis to the expression xy formed by Composition we find that no condition of interpretability is involved.

To apply a similar principle of analysis to the expression $\frac{x}{y}$ formed by Abstraction we must observe that as Abstraction is only intelligible by reference to Composition the expression $\frac{x}{y}$ only signifies a class which by Composition with y gives x. Let w represent such a class.

Then $\qquad x = yw \qquad$ ( ).

Now if we compound both members with the class $(1-y)$ attending to the condition $y(1-y) = 0$ we have

$$x(1-y) = 0$$

Logically interpreted this demands the non-existence of the class whose members belong to the class x and not to the class y. And this agrees with the condition assigned in ( ). Interpreted in dual Algebra it permits the combinations

$$x = 1 \qquad\qquad y = 1$$
$$x = 0 \qquad\qquad y = 1$$
$$x = 0 \qquad\qquad y = 0$$

and excludes the combination

$$x = 1 \qquad\qquad y = 0 \qquad\qquad [C 42.23]$$

Now the substitution of the three former combinations in the expression $\frac{x}{y}$ gives results which belong or may belong to the dual algebra viz. 1, 0, $\frac{0}{0}$ the last of these results admitting of either of the values 1 and 0. The substitution however of the final combination $x = 1$, $y = 0$ in the same expression leads to a result which as representing infinity is not included in the dual Algebra.

The deduction of all three conditions is however more readily and in the last case more directly performed by the process of development hereafter to be described a process essentially founded on that peculiar law of thought which forms the basis of the previous deductions.

The general conclusions to which these investigations lead may now be collected into the following summary.

Logical propositions being expressed by a system of notation in which 1st the Universe is represented by the symbol 1 and Nothing by the symbol 0 2ndly Classes of things by letters 3rdly The elementary operations of Conception viz. Addition Subtraction Composition and Abstraction by the corresponding signs + − × ÷ 4thly the relation of objective identity by the symbol =. Then will both the formal laws of thought and operation and the formal conditions of interpretability in that system agree with the corresponding laws and conditions of a dual Algebra recognising no other special symbols of number than 1 and 0. [C 42.24(+25)][18]

It appears then that there exists a perfect formal identity between Logic represented by symbols in the scheme above explained and the dual Algebra whose nature and character have been described. Upon this identity the methods developed in the Laws of Thought are founded. I have not however in that treatise so fully considered the grounds of the relation upon which its methods rest as I have done in the previous sections of this paper. The identity of the formal laws of operation was demonstrated but not the fact that the formal conditions of interpretability are the same also and that these conditions are a necessary consequence of the formal laws. And in accordance with that first philosophy of reasoning of which all symbolical methods are but special applications I conceive that the proof of the formal identity of the conditions of intelligibility in the two systems was not required. It suffices that the laws of operation are the same and that the results to which we are led by the methods of the dual Algebra are interpretable in Logic. Still I apprehend that the identity of the systems as respects not only the formal laws of operation but also the formal conditions of interpretability is a fact of great moment and significance and that it would be [C 42.26] unphilosophical to regard it as a merely accidental coincidence. It has unquestionably a deep and real foundation in the constitution of the human mind.

Two questions of great importance here present themselves.

First. Is it necessary to the mathematical development of Logic that we should take into account its formal relation to the dual Algebra?

To this I reply that it is certainly not necessary. We might for instance instead of representing Universe by 1 and Nothing by 0 have employed other symbols e.g. we might have represented Universe by u and Nothing by n. Only had we done this the symbols u and n would have required in virtue of the formal laws of Conception to be used according to the same rules as the symbols 1 and 0 in the dual Algebra.

Secondly. What then is the advantage of employing the dual Algebra?

The advantage I conceive to be that when we have established by a just analysis of the intellectual operations our right to the employment of a system whose laws are in every respect coincident with those of Logic but whose ideas are [C 42.27] more simple we are led by easier steps of suggestion to the constructive development of the science. I say *a system whose ideas are more simple* because it is I think more readily perceived that $x \times 1 = x$ when x represents a number, 1 unity and the symbol × the operation of multiplication than when x represents a class 1 the Universe and the sign × the operation of Composition. Once however that the method is fully developed it is of no importance whether we consider its processes as founded immediately upon the laws of thought in Logic or only mediately thereupon through the intervention of the dual Algebra.

The most remarkable difference between Logic and dual Algebra a difference however which is material and not formal is that their spheres of interpretation are not merely different but not even connected by a perfect correspondence and analogy. In the dual Algebra any general symbol x admits only of the special and definite meanings 0 and 1 but in Logic such general symbols admit of other special and definite interpretations than Nothing and Universe even of any class interpretation whatever i.e. of any meaning which can be expressed in language by a general name. The symbol x may be determined to mean Nothing or Universe but it may also be de- [C42.28] termined to mean "men" "golden mountains" "Utopia". But under all its diversities of meaning it remains subject to a common system of formal laws and these alone affect the processes of inference.

Analysis of Reasoning resumed

The distinction of reasoning as synthetical and analytical has been already noticed. As the synthetical form consists in the direct application of the laws of thought with or without the aid of a symbolical language to the premises and as those laws have already been investigated there is little more to be said upon the subject. It remains then to explain more fully the nature of the analytical method of reasoning and to shew how it is developed by means of symbolical forms.

It must be remarked then that this method always has reference to a proposed end. It begins with some such question as the following. What unknown concept do we seek explicitly to determine and by means of what other concepts is its explicit determination sought? I say *explicit determination* because it is the office of a conclusion not to present to us new truth but only to bring into an explicit form and statement some portion of that truth which was implicitly involved in the premises. And having determined what kind of conclusion is sought we proceed 2ndly [C42.29] to express the most general form which that conclusion in accordance with the necessary laws of Judgment and in perfect independence of the particular information conveyed in the premises must have. This being done we proceed 3rdly to determine by means of the premises whatever is arbitrary in the general form so far as such determination is possible. It is in the last step chiefly that the principle of substitution is employed.

The two orders of procedure which have been described viz. the synthetical and the analytical are clearly inverse to each other and they rest upon fundamentally different but perfectly consistent views of the nature of a conclusion. For we may regard a conclusion either 1st as a consequence following directly from the premises or 2ndly as a condition necessary in order that the premises may be thought as true. There is no difference in point of validity between the methods. In comparing them we should say that the method of synthesis is the more simple in theory that it is necessarily ostensive not demanding even a recognition of the formal character of the processes of inference – to counterbalance all which advantages it must be added that it is necessarily limited in application. The method of analysis on the other hand rests upon a deeper basis of theory, it regards the processes of [C 42.30] inference essentially formal, it is perfectly general in its application.

What now is the most general problem which Formal Logic can propose? In considering this question I shall confine myself to the class of propositions termed categorical. It will suffice to do this because the theory of hypothetical propositions is formally identical with that of categorical propositions. With this restriction then the premises of any train of argument consist of propositions each of which expresses a relation between two concepts. Each again of the two concepts thus connected is either an elementary concept expressed by a single name or description or it is formed by the combination of elementary concepts according to the laws of the faculty of Conception already investigated. Ultimately therefore we may say that the premises of an argument consist of propositions expressing relations among elementary concepts.

The conclusion which we seek must also be a proposition and express a relation connecting all or some of the elementary concepts involved in the premises. If it involve only some of them the others must have been eliminated. And hence an essential part of a general method in Logic must be a process of elimination i.e. a process by which we may get rid of those concepts which we do not wish to retain in the [C 42.31] conclusion and which we may consider as only deserving of consideration in so far as they help to establish relations among the other concepts.

Again it may be required to determine the conclusion either 1st as a direct expression of existence or non-existence or 2ndly as an expression of the relation in which a given concept whether elementary or formed by a combination stands to the other concepts retained. For example if the premises expressed certain relations between the concepts "coals" "minerals" and any other concepts and if it were required to determine the explicit relation thereby established between coal and minerals a definite process of inference might lead to that conclusion in the form,

"Coals that are not minerals do not exist"

and another process of inference or the same carried still farther might lead to an equivalent conclusion in the form

"Coals are minerals"

and this latter might be arrived at as an answer to the question "How far is it possible to define the object *coals* by means of the concept *minerals*" Of these forms of the conclusion the former is simply an expression of non-existence the latter is an expression of the relation in which a given concept stands to another concept. [C 42.32]

The general problem of Formal Logic in reference at least to categorical propositions is therefore the following.

Given any set of propositions expressing logical relations among concepts required to express by a single proposition the whole of the relation connecting either 1st all of these concepts or 2ndly any of them chosen at liberty. Required moreover to *express* such relation either 1st, as an expression of existence or non-existence or 2ndly as an expression of the relation in which some given subject formed by the elementary concepts stands to the other concepts.

This problem under its most general form, the symbolical calculus enables us to solve. The formal laws whose naked expression has been given in the previous section furnish us with the requisite methods. For the fundamental processes which these methods involve viz. Development Elimination and Reduction I must refer to the Laws of Thought. But I will say a few words here on the philosophy of the most important of these processes viz. that of Development more especially as it is in this process that the analytical element of reasoning finds expression.

Resuming the example just considered in which from supposed premises it was required to express "coals" in terms of "minerals" we may remark that it is possible antecedently to [C 42.33] any knowledge of the meaning of the terms or to any information conveyed in the premises to posit the necessary proposition

"Coals are either minerals or not-minerals"

If there were another term required to be taken into account e.g. "vegetable origin" we might quite as independently form the necessary proposition

"Coals are   either minerals of vegetable origin
              or not-minerals of vegetable origin
              or minerals not of vegetable origin
              or not-minerals not of vegetable origin."

The only effect of a knowledge of the premises would be to enable us to limit in some more definite manner these necessary propositions. They might for example enable us to reject in the first form of conclusion the alternative "not minerals" and to affirm "Coals are minerals". They might possibly enable us to say "Coals are all minerals". And in like manner the four alternatives involved in the second form of the conclusion might severally be affected or placed under different categories of thought by means of the premises. All such modifications presuppose the existence of necessary and *a priori* forms of *Judgment*. [C 42.34]

He who in ordinary reasoning adopts this mode reasons analytically. He begins with a necessary proposition and imposes upon it special limitations so as to make it accord with the premises.

Now the method of development furnished by the symbolical calculus is in reality but the same process presented in a purely scientific form and in immediate connexion with a dependence upon the laws of thought. It virtually begins with presenting the conclusion in the form of a necessary proposition but it derives that necessary proposition from the fundamental law of thought expressed by the equation

$$x(1-x) = 0$$

It proceeds by a general method to determine what I have called above the categories or general relations under which the several terms or elements involved in the necessary proposition must be contemplated by the mind in accordance with the premises. This determination proceeds essentially upon the principle that Logic is formal in its character – that those relations do not depend upon the meaning or nature of the concepts involved in the premises but only upon the relation in which they stand. I will endeavour to illus- [C 42.35] trate these positions by one or two simple examples first giving the symbolical solution according to the method of dual Algebra in the Laws of Thought and then interpreting the several steps of the solution.

Suppose that from the proposition "Men = rational animals" it were required to find explicitly a definition of "rational beings" in terms of "men" and "animals".

If we represent the concept "men" by x "rational beings" by y and "animals" by z we have the equation,

$$x = yz \tag{1}$$

Hence
$$y = \tfrac{x}{z} \tag{2}$$

and developing the second member

$$y = 1zx + 0z(1-x) + \tfrac{1}{0}(1-z)x + \tfrac{0}{0}(1-z)(1-x) \tag{3}$$

The interpretation of which is the following:

> 1st Rational beings consist of all animals that are men, no animals that are not men and an indefinite remainder (some none or all) of beings that are neither animals nor men.

> 2ndly Men that are not animals do not exist.

I will first make a few observations upon the symbolical equations (1) (2) (3) in order.

In (1) which is the symbolical expression of the [C42.36] premiss the concept y whose explicit definition is sought appears in *composition* with z. From this connexion it is freed by the inverse operation of Abstraction. The equation (2) expresses this fact and shews that it is by the abstraction of the concept z from the concept x that the definition of y must be obtained. Equation (3) exhibits the result of the abstraction as obtained by the process of development that process depending not upon the meaning of the symbols z, x, but only upon their formal laws. We may distinguish in the resulting expression for y two classes of elements, viz.: 1st the terms zx, $z(1-x)$, $(1-z)x$, $(1-z)(1-x)$ 2ndly the coefficients 1, 0, $\frac{1}{0}$, $\frac{0}{0}$ with which those terms are affected. The former class are quite independent of the manner in which z and x enter into the original equation (1). The explicit definition of the logical concept y in terms of the logical concepts x and z, whatever the given relation among those concepts may be, must in its ultimate state of resolution involve those elements not because the concepts $x\,y\,z$ are logical but in virtue of the law of thought $x(1-x) = 0$.[19] The second class of elements viz. the coefficients are however dependent upon the premiss. They shew under what logical categories the several terms of the neces- [C42.37] sary proposition must be thought in connexion with the premises. Let us consider these coefficients separately.

The coefficient 1 attached to the concept zx implies that the whole of the class represented by that concept is signified (Laws of Thought p. 92).

The coefficient 0 attached to the concept $z(1-x)$ indicates that *none* of the individuals included under that concept are signified.

The coefficient $\frac{1}{0}$ attached to the concept $(1-z)x$ indicates not only that no part of the class represented by that concept is included but that the existence of such a class is forbidden by the premises.

The coefficient $\frac{0}{0}$ attached to the concept $(1-z)(1-x)$ implies that an indefinite portion of the class represented by that concept is signified.

If we endeavour to interpret the coefficients by independent concepts we shall have

1 = Universe or Things existent*

0 = Nothing or Things non-existent

$\frac{1}{0}$ = Things impossible

$\frac{0}{0}$ = Things indefinite

* Had the principle of analogical interpretation been admissible we should probably have been led to represent the concept *Universe* by the arithmetical symbol for infinity and not by the symbol for unity. Formal law is indeed the only element of analogy which it is permitted to recognise in the determination of the [illegible] of symbols.[20]

Respecting the ground of the two last interpretations let it be observed that $\frac{1}{0}$ would represent a class of things such [C42.38] that if we seek by Composition the class common to that class and to Nothing we shall obtain the class All-things ( ) now no such class exists or is conceivable. Retaining the language which is grounded upon the notion of Class we must interpret the coefficient $\frac{1}{0}$ by Things impossible. Again the expression $\frac{0}{0}$ represents a class of things such that if we compound it with Nothing we obtain Nothing. But this condition is answered by all concepts whatever. Hence $\frac{0}{0}$ must be interpreted as things indefinite or unlimited.

Considered then in the concrete the four coefficients of the development admit of the interpretation Universe Nothing Things impossible Things indefinite. Considered with reference to the abstract notions which they express we have the categories 1st Existence or Totality 2nd Non-Existence or Nothingness, 3rd Impossibility, 4th Indefiniteness. It is a remarkable circumstance that however numerous the terms of a conclusion those terms appear under no other relations or categories than the above. The categories do not however necessarily all present themselves in every conclusion nor in fact is there any other general observation to be noted than that the above are all which [C42.39] do appear. We may therefore with propriety term them logical categories. A similar system with corresponding interpretations results from the application of the method of development to hypothetical propositions.*

I will endeavour to illustrate the above process, and subsequently that of elimination, without the introduction of mathematical forms. In both these processes much use is made of the operation of Composition. It is to be remembered that the Composition of two concepts means the forming of the concept of a class of things consisting of the individuals which are common to the classes represented by the concepts compounded and hence "Universe" being the concept of the class which contains all existing things "Nothing" of the class no things, that

Universe in composition with any concept = that concept   [C42.40]
Nothing in Composition with any concept = Nothing
Any concept with itself = the same concept

And in particular

---

* I have retained in the present paper the familiar division of propositions into categorical and hypothetical a division which belongs rather to the form of expression than to the matter of propositions. The true ground of distinction which however is in its practical results nearly coincident with the above will be found in the Laws of Thought Chpt. XI.

Universe in Composition with Universe = Universe
Universe with Nothing = Nothing
Nothing with Nothing = Nothing

Now in the example above considered the premiss is

"Men = rational animals"

and it is required to determine "rational beings" by means of the concepts "Men" "Animals".

Supposing ourselves ignorant both of the premiss and of the meaning of the terms "Men", "Animals" we may at once by the law of necessary Judgment say

I   Rational Beings =
        Men animals in composition with
            some unknown concept A,
        Men not animals in composition with
            some unknown concept B,
        Animals not men in composition with
            some unknown concept C
        and not animals not men in composition with
            some unknown concept D.

The premiss presents to us "rational beings" simply as a class which by composition with "animals" gives "men". And hence by the principle of substitution [C 42.41]

II That class which by composition with
    animals gives men =
        Men animals in composition with
            some unknown concept A,
        Men not animals in composition with
            some unknown concept B,
        animals not men in composition with
            some unknown concept C,
        and not animals not men in composition with
            some unknown concept D.

And the principle upon which the categories are to be determined is that we may in the above proposition replace the special terms "men" "animals" by the concepts "Universe" "Nothing". The proposition II then becomes.

III That class which by composition with
    Universe gives Universe =
        Universe in composition with the concept A
        all the terms following this becoming Nothing.

Now Universe is the only concept which by composition with Universe gives Universe. And Universe in composition with the concept A simply gives the concept A. Hence the concept A is determined as Universe, so that referring to the necessary proposition I we see that Rational beings will include all men-animals.

To determine the concept B replace Men by Universe animals by Nothing whence not men by Nothing and not animals by universe. We find from I [C 42.42]

> That class which by Composition with
> Nothing gives Universe =
> Universe in composition with the concept B.

Now that there should exist any class which by composition with Nothing should give Universe is impossible. The concept B therefore must represent Things impossible. Hence the concept men not animals appears in the conclusion under the category of impossibility.

To determine the concept C let "men" be replaced by Nothing "animals" by Universe. We find from I

> That class which by composition with
> Universe gives 0 =
> Universe in composition with the concept C.

But the class which by composition with Universe gives Nothing is Nothing. Hence the concept C is Nothing. Whence animals not men appear in the conclusion under the category of non-existence.

Lastly to find the concept C let "men" and animals each be replaced by Nothing. We have from I

> That class which by composition with Nothing
> gives Nothing =
> Universe in composition with the concept D.

Now any class in composition with Nothing gives Nothing. Hence the concept D is any class whatever. The class of not-men not-animals appears then in the conclu- [C 42.43] sion under the category of indefiniteness.

Collecting these results together we have

> Rational beings =
>   All men-animals,
>   no animals not-men
>   an indefinite remainder (some none or all)
>   not-animals not men;

the existence of men-not-animals being at the same time declared impossible in virtue of the premises.

Cumbrous as this method will doubtless [appear] it may serve to explain the nature of a logical conclusion as formed by the limitations of a necessary proposition, and also to shew in what way the two limiting concepts of Nothing and Universe may be employed to effect the necessary reduction.

But this employment would scarcely occur to any mind unacquainted with the researches upon which the method is founded nor can it be employed with even tolerable convenience except in connexion with a symbolical calculus.

Much more remarkable is the method of elimination. It reduces to a single rule or principle the elimination of any concept from any proposition however complex or from any system of propositions and the rule though perfectly general is such that it would seem in the highest degree improbable that it should have been discovered without an expli- [C 42.44] cit study of the formal laws upon which it is founded. I will state the rule as applicable to single propositions and briefly illustrate it by an example.

The law is the following If any proposition be reduced to a form in which it *expresses* the non-existence of a class, the elimination of any concept involved in the *expression* of that class will be effected by substituting for that concept in the given expression the limiting concepts "Universe" "Nothing" successively compounding the expressions thus obtained and affirming by a proposition that the class so defined is non-existent.

Let us apply the rule without intervention of mathematics to the proposition "Men = rational beings that are animals" and suppose it required to eliminate the concept rational beings. We have

Men except rational beings that are animals do not exist,

in which form it expresses the non-existence of a class. In that class substituting for "rational beings" "Universe" we obtain "Men except animals", substituting for "rational beings" "Nothing" we obtain "Men" compounding these results we have the expression "Men except men that are animals" or [C 42.45] "Men that are not animals" and lastly affirming by a proposition the non-existence of the class so defined

"Men that are not animals do not exist"

a proposition in which the concept "rational" no longer appears. The method of development applied to this result would bring it to the form

"Men are animals"

Lastly I will apply this rule in combination with reduction to the premises of the syllogism considered viz.:

Men are mortal
Kings are men

These may be reduced to propositions expressive of non-existence in the form

Men not-mortal exist not
Kings not-men exist not

And hence we have the single proposition

Men not mortal and kings not men exist not.

I stop to remark that this proposition is really and truly a single proposition. We can form the concept of a single group or collection of individuals composed of two parts viz. "Kings not men" and "Men not mortal" these parts being from the very form of expression mutually exclusive and therefore admitting of addition. And of this collective group [C42.46] thus formed we can affirm non-existence. It is in the oneness of this affirmation that the singleness of the proposition consists. The reduction of complex systems of propositions to single propositions constitutes one of the three great processes of the calculus of Logic – the converse resolution of single propositions into systems of propositions being a particular application of the method of development.

In the class whose non-existence is declared let us substitute for "men" the concepts "Universe" "Nothing" in succession, the results are "things not mortal" "Kings". Hence, compounding these and affirming by a proposition the non-existence of the resulting class, "Kings not mortal exist not" a proposition which development would reduce to the form "Kings are mortal".

It will be noted that in the above applications reductions of various kinds are employed of which no account has been given e.g. the reduction of the proposition "Men are mortal", to the form, "Men not mortal exist not". In the symbolical methods no such adventitious [aid] is required or admitted all transformations whatever being resolved into applications of the three general methods of Development Elimination and Reduction the basis of which consists so entirely in [C42.47] the *formal* laws of thought that these methods are the same without any difference in the dual Algebra whose formal identity with Logic has been demonstrated. [C42.48]

Chapter VI

[Preparatory Notes]

—————————————[Between 1854 and 1856]—————————————

Of the nature and scope of Logic as considered a Science

It is a very imperfect definition of Logic to say that it is the Art of Reasoning or that it is the Science or even that it is the Science and Art of reasoning. It is a juster and more complete description of it to say that it is the Science of Thought. For it is impossible to give any satisfactory account of the process of Reasoning without first analysing other processes of Thought, Conception, and Judgment. Nor does it suffice merely to recognize the existence of such processes and to describe their objects. We must also investigate their laws. Except upon the ground of such investigation any attempt to analyse the process of Reasoning must fail. It must sink into little more than a mere classification of the forms of inference.

We cannot divide the domain of Thought into its several constituent provinces and establish a perfect mastery over any one of them without taking possession of the others also.

The true definition of Logic as I conceive is the following. Logic is the Science of Thought as exercised upon things and as expressed in Language. [A92.11][1]

Logic is the Science of the Laws of Thought as consisting of the operations of Conception Judgment and Reasoning.[2]

Definitions of Conception, Judgment and Reasoning

[Conception]

Conception is that power by which we mentally picture or represent things to ourselves.[3]

The origin of the materials upon which conception operates is experience.

Experience is either by means of the senses or by means of internal consciousness, the former the channels by which our knowledge of the qualities of matter in all its modifications is derived the latter the medium by which we are made acquainted with our own moral nature – with the idea of beauty with the notions of power, of causation etc. Locke, Language.

Resemblance between our conceptions and the things for which they stand [is] not necessary for reasoning. In the latter process conceptions degenerate by abstraction into mere signs.

But as signs they are still subject to laws.

A sign is therefore a conception of which that portion derived from the senses or experience is as it were suspended, and the formal laws remain.

The formation of signs is therefore gradual, it begins with experience it is completed by the power of abstraction by which we can fix upon those qualities which are common to many [A92.1] individuals, partly by that power of substitution or representation by which we can make one thing stand for another with which it has been associated and indifferently for any one of a series of others with each of which it has been associated.

Hence signs are representative in form. Their connection with the thing signified is by the law of association but not as founded in resemblance. Hence they are arbitrary in form.

For the purpose of reasoning their laws must be understood, and their interpretation must be known.

To what extent can the content of a sign be suspended? For this All imaginations of qualities, impressions etc. physical or moral may be suspended – but the distinction between the signs of different things must be maintained.

Analysis of conceptions[4]

1st. Pictorial or representative element derived from imagination.
2nd. Formal character or relation to law.

The former not necessary for the purpose of reasoning but capable of suspension. Important for judgment.

One reason not why it is not necessary but why it cannot be employed is the imperfection of the senses, hence of our perceptions – hence of our conceptions viewed as belonging to the imagination. [A92.2]

Different Theories of Abstraction.

1st. That we form a general notion not accurately representing any particular individual.

2nd Berkeley's theory that we form the conception of an individual and make this stand for the genus.

Both these theories contain an element of truth. The true doctrine is to be found in a consideration of the imperfection and vagueness of all sensual images whether as immediately given in perception or as reproduced by imagination. All that is needful is that they should serve the office of signs, should suffice to denote what we mean. Particular measures [are] unimportant.

General Question: How are the laws of thought to be determined?

Particular Question. How are they to be determined with relation to the particular faculty of conception.

1. By considering the different operations involved in conception – 2ndly. By enquiring under what conditions these operations are possible.

Operations of Conception

The operations involved in conception are the following:

Direct	Inverse
Addition	Subtraction or Removal
Combination	Decomposition

The operation of addition and its inverse is the basis of extension. Combination and its inverse – of comprehension. [A92.3]

Symbols must be employed both for conceptions and the operations by which conceptions are modified – also for the forms of judgments.

The laws of the operations will determine the laws of the symbols.

Judgment

The element of judgment is the proposition.

Two theories of the nature of judgment.

1st. That of predication or inclusion.
2nd. That of identity.
The latter involves the doctrine of the quantification of the predicate.
Extent of this quantification determined by the laws of conception. It must not exceed the bounds of Logic.
Hence numerically definite propositions are excluded.
We have: 1st. the Universal 2ndly the indefinite (v) 3rdly 0. The latter excluded.
In pure Logic All men are sinners means All men that exist in the universe of discourse are sinners. A connection is established between categoricals and hypotheticals.
From the principle of the expression of identity – the distinction of quantity – and the introduction of privative concepts all the forms of judgments are determined *a priori*.
Origin of quantification in composition with the concepts Universe, Nothing and the concept Indefinite with the undefined and the limits of the defined. [A92.4]

Classification of these Forms

Objection respecting the negative (that it constitutes a denial). Solution of the difficulty by means of hypothetical propositions.

Necessary judgments derived from the laws of thought (trifling propositions).

Kant's distinction of analytical and synthetical judgment rather necessary and contingent or necessary and empirical.

Reasoning

Primary division into synthetical and analytical.

Synthetical recognizes direct operations only.

Its fundamental laws are 1st. The law of substitution. 2nd. Like operations performed on like subjects produce like effects.

The law of substitution is the principle of identity continued.

The law of operations is analogous to the law of causation.

Analytical reasoning has for its object the expression of a conception in terms of given elements by means of other conceptions with which it is implicitly connected by means of the [A 92.5] premises the discovery of the manner in which the form of a concept whose matter is given is effected by given premises.
It seeks to determine a notion 1st By those general principles contradiction and excluded middle which prescribe to a certain extent beforehand the forms of all conceptions. 2ndly. by imposing thereupon the further limitations involved in the premises.
Thus Analysis proposes to itself a given end.

Further it is by Analysis that we must proceed when the inverse operations of Conception are involved.

The peculiar principle thus introduced is that where like operations have produced like results the subject upon which those operations have been performed must be included in the most general determination of the subject, upon which the other operation has been performed.

Or the two laws may be thus stated. Where operations are definite like operations performed upon like subjects produce like results.
Where operations are indefinite like operations performed upon like subjects produce results whose range of possibility is the same.

This principle has also a metaphysical analogue referring not to the calculation of effects but to the research of causes or of antecedent states e.g. where two like causes of change have produced like results we may conclude not that the antecedent state of things upon which the one cause operated was [A 92.6] the same as that upon which the other cause operated – but that it must have been equivalent to some one of the possible states of things out of which the other cause might have elicited the result observed.

[Further Considerations][5]

The question now arises, how is it possible to determine scientifically the result of an inverse operation. In other words How shall we determine the most general expression of the subject upon which the performance of a given direct operation shall produce a given definite result.

The answer to this question is contained in the statement of another general principle governing the connection of Thought with Language viz. the processes of Reasoning depend only on the formal laws and are independent of the interpretation of the symbols employed.

Antecedently to the publication of the Laws of Thought no solution of this question had been attempted nor had the question itself been distinctly conceived or entertained.

The solution given in the Laws of Thought is a complete one, but seems to introduce an extra logical element, viz. the analogy which exists between the formal Laws of Logic and the formal laws of a particular species of Arithmetic.

But it is one thing to give interpretation to a principle or rule within the bounds of pure Logic and another [A92.7] thing to be able to demonstrate its truth upon purely logical grounds.[6]

It is not likely that the purely logical law here to be developed by interpretation from a law founded upon the relations of Logic and Arithmetic would have been suggested by purely logical considerations.

In an attempt to determine a priori the cause and development of a science we are very apt to suppose that it must be kept pure from all admixture of elements from without. And yet instances are continually arising which teach us that there is a vital connection among the different sciences.

Statement of this analogy and of the calculus founded thereupon.

Can the processes be explained and interpreted without introducing the arithmetical notions? It can.

The two limits of conception are Universe and Nothing.

Each of these concepts is subject to the formal laws of concepts in general and to peculiar laws dependent upon its own interpretation.

But the *laws which are common to the two concepts are the laws of all concepts whatever*. This is the first property to be noted.

Again each of these concepts by virtue of its peculiar formal laws when entering into composition with other concepts is merely determinative of quantity. Second Property. [A92.8]

The principle of this employment in analysis is the following: If any proposition is equally true when for a given concept entering into its expression we substitute Universe and Nothing it is true universally.

Or this We may conduct all formal processes of reasoning on the supposition that concepts admit of no other interpretations than Universe and Nothing.

An illustration   from development.
     do.   from elimination.

Origin of the four categories of Judgment existence, non-existence, indeterminateness, inconceivability.

Objection that we do not actually reason thus.

Reply It is a mistake to suppose that the actual performances of our nature in any case fully answer to its faculties and capacities.

We are in all things constituted with reference to an ideal standard.

2nd objection. The concept Universe limits the individuals comprised under it by no quality but that of existence. Now we cannot conceive of existence merely apart from qualities.

Reply. Here as in perhaps all purely scientific inquiries we have to do with limits. We cannot abstract every quality and contemplate pure existence – but this is the limit to which abstraction tends – and it is in accordance with the philosophy of geometry etc. that we make the limit the subject [A92.9] of our formal investigations.

Universe and Nothing regulative concepts.

Deep in the ground of all our reasoning about particular things there is a reference to the universal. We may not be conscious of it but it is not the less a part of the very constitution of the mind. [A92.10]

Chapter VII

General Summary

────────────────[Later than 1854]────────────────

Theory of Formal Logic

A. The power of conceiving any collection of things possessing some common attribute as constituting a class being pre-supposed there are two distinct ways in which simple conceptions thus furnished may by the mind's own activity be made complex. 1st. By aggregating the *members*. 2ndly. By combining the *qualities* of different *classes*.

B. The mind can proceed by courses respectively inverse to the above so that there are in the whole four kinds of operations.

C. Thus – 1st – Given a conception x and another conception y we can form a new conception z by aggregating y to x so that $x+y = z$. Inversely, given a conception z and a conception y, we can propose the question what is that conception x with which if we aggregate y we shall obtain the conception z. 2ndly. Given a conception x and another y, we can combine these into z so that $xy = z$. Inversely given y and z we can inquire what is that conception x which if y be combined with it will produce z.

D. The *primary* operations are those of aggregation and combination. And it may be questioned whether these are not the *only* operations connecting simple and complex conception. In their direct performance we pass from a subject (simple) conception to a resultant (complex) conception and we so pass by means of another conception which we either aggregate or combine with the given simple conception. In the inverse [C59.1] procedure we have to seek what is the subject conception with which such a proposed aggregation or combination must be effected in order to produce a given complex conception as the result, so that the *operation* is the same whether we proceed in the one direction or in the other, only in the one case we ask what is the *result* in the other what is the subject.

E. Even in the case of *segregation* which seems at first sight to be independently possible it would seem that the direct operation of aggregation is presupposed. From the conception "human race", by segregation of "females" we arrive at that of males, but it may be doubted whether the very possibility of this does not rest upon antecedent perception that human race is a whole formed by the aggregation of the parts "males", "females".

F. And whether this be so or not the two operations above defined as inverse are fully *determined* by the inverse definition i.e. It is sufficient to determine the result of the segregation of a part from a whole to say what is that part with which if the given part be aggregated the given whole will be obtained.

G. The operations of aggregation and combination are subject to formal laws which find one expression in the known logical principles of contradiction and excluded middle.

G'. If we express conceptions by letters, aggregation by + the symbol of arithmetical addition, combination by juxtaposition of letters like arithmetical multiplication then the laws of the two direct operations of aggregation and combination are formally the same as the laws of addition and multiplication in an arithmetic of 0 and 1. [C 59.2]

H. Hence by F, the laws of the inverse processes if expressed by the same notion − ÷ as in the arithmetic of 0 and 1 will be subject to the same formal laws.

I. Again propositions as implied by the very meaning of the copula *is* are expressions of the identity of the *things* denoted by the subject with the things denoted by the predicate − each term under the implied limitations as to *quantity*.

J. And if this identity be expressed by the arithmetical symbol = axioms will exist which will lead to formal processes in equations identical with those which exist in the arithmetic of 0 and 1.

J'. From G and J flows the symbolic method developed in the Laws of Thought.

K. But now arise these questions: 1st. Can that symbolic method be established independently of the formal analogy above noticed and without transgressing the boundaries of pure logic. 2ndly. If they can, what is the logical doctrine or theory upon which it rests, 3rdly. Is there an essential ground of the formal analogy noticed?

With these may be associated another question viz. What is the general *result* of the method?

These questions will now be considered.

L. The laws of the mental operation of conception as expressed by a general symbol x are the same as the laws of a general symbol x denoting either 0 or 1 in Arithmetic (G). And the laws of the particular symbol 0 denoting Nothing and of the particular symbol 1 denoting Universe are the same as the laws of the particular symbols 0 and 1 respectively in that Arithmetic. Hence the laws of the literal symbol x in Logic are the same as the laws which are *common* to the two [C59.3] symbols 0 and 1 in Logic. Or to quit symbolic language The formal laws of the terms or marks by which we express our general conceptions of things are identical with the formal laws which are common to the terms or marks by which we express the particular conceptions of Nothing and Universe.

M. Now conceptions may according to the familiar language of logicians be contemplated either by *extension* or by *intension*. As adopting this distinction that which characterises white things is whiteness red things redness and so on so that which might be said to characterise Universe contemplated as a class is existence and that which characterizes Nothing is non-existence. Thus if we asked what is the whole of that collection of things of which existence alone is predicated and required we should [say] it is the Universe for that is by definition the aggregate of existing things. If we ask what is that collection of things of which non-existence is predicated or demanded we should say There is not such connection of things – it is Nothing.

N. Hence the formal laws of things as defined intensively by the possession of quality are the same as the formal laws of things considered merely under the qualitative notions of *existence* and *non-existence*.

This conclusion may be independently confirmed.

O. It will at once be admitted that reasoning is so far formal that its processes do not depend upon the particular meaning of the class terms employed. We employ the same *forms* whether we were reasoning white things or black things or things of any colour or quality whatever and hence we may at once infer that we might make abstraction of all such particular [C59.4] manifestations and ascend to that which is left behind – viz. mere existence or non-existence. These are the only abstract ideas. The necessity of taking account of non-existence as well as of existence if not a first sight evident becomes so on reflection. For the things mentioned in the premises of an argument may from the very connection thereby established among them be some of them or some combination non-existent.

P. From the formal laws of conception flow the so-called principles of contradiction and excluded middle. In virtue of these we may say of any existing class of things that it either 1st possesses any assumed *arbitrary* quality or 2nd does not possess that quality – that if two arbitrary qualities are assumed the proposed class possesses either 1st possesses both or 2ndly possesses the first and not the second or 3rdly possesses the second and not the first or 4thly possesses neither and so on. Such is the origin of what a large portion of what are called necessary judgments viz. of those which enable us to predicate the possible alternatives under which any class of things whatever may be spoken of with reference to any quality or qualities whatsoever.

Q. The general form under which the solution of logical problems is given in the Laws of Thought shews that every *conclusion* consists of a necessary judgment of the above description (and thus far independent of the premises) made to a certain extent determinate by means of the premises – the determination consisting in placing each term of the alternatives in the predicate under some one of the four following categories viz. [1st.] universality, indicating that the whole of the [C 59.5] individuals expressed by terms are included in the *subject* 2ndly. negation indicating that none of those individuals are included in the subject 3rdly. potentiality indicating that some, all or none of those individuals are contained in the subject 4thly. impossibility indicating that the class of things denoted by the term exists nowhere either in the subject or out of it.

R. The *categories* originate from the two fundamental conceptions of existence and non-existence and the operations by which conceptions are combined.

S. The ultimate theory of reasoning in reference to categorical propositions may then be summed up as follows. By an intellectual operation we ascend from the presentations of the senses to the conceptions of things under the relations of genus and species which are *expressed* by the general terms of language. These conceptions we have the power of combining by certain definite operations subject to definite laws, so as to form new conceptions. These conceptions we have also the power of connecting in thought under the relations expressed by propositions. Such relations are either necessary i. e. founded in the very laws of thought or empirical. When any empirical propositions are given the object of reasoning is to deduce from them other propositions connecting the same conceptions or some of those conceptions and the object of a theory of reasoning is to shew 1st. how this is done 2ndly. what is the general character of the result. The answer to the first question is that to ascertain very limited extent we can do so by direct processes consisting in the application of such axioms as "Things identical with the same [C 59.5a] thing are identical with each other", If to equal things equal things are added or if from equal things equal things are taken or if of equal things [equal things] are predicated, the resulting things are equal – but the only general method consists in this viz. that the conclusion is expressed first in the form of a necessary proposition in virtue of those formal laws of thought which constitute the basis of the logical principles of contradiction and excluded middle and then the several alternatives in the predicate term determined under the four categories above described by virtue of the principle that every conception involved in the premises is the conception of a class of things which is either existent or non-existent and that the formal laws of such conception are exactly fulfilled by causing it [to] merge in succession into the two elementary conceptions of existence and non-existence.

T. Reverting to K it must now be considered whether there is any essential ground of the formal analogy between the conceptions of non-existence and existence in Logic and of 0 and 1 in Arithmetic.

The answer is that the idea of unity seems inseparable from that of a *class* of things contemplated as existing. From the conception of any class of things let us abstract all that is distinctive in point of quality without at the same time ceasing to regard it as a class and let us examine what ideas remain and what is their logical import.

1st. There is the metaphysical notion of substance – but that does not at all affect our logical processes. Some reject it wholly and say that we have nothing to do but with qualities, others retain it but as a mere substratum – even as the canvas of the painter according to its design at least supports [C59.6] the colours which he lays on but does not affect the laws of their harmonious blending. 2ndly. There *may* be the idea of multitude in reference to the component individuals of the class – but the formal reasonings of logic do not in [any] way depend upon the *number* [of] the individuals contained in the class. 3rdly. There is the idea that the class of things is *one* class – one as an object of thought. And this idea seems to be indispensable. We have said however that the connection of terms implied by the premises may indicate that some of the things are non-existent. Thus the ideas of existence and non-existence must both be retained – as essential even in purely logical considerations.

As a proposition is in its essence something which is true or false so a class of things is in its essence something which is existent or non-existent.

U. The idea or conception of Class is anterior to that of Number. The latter presupposes the former. If we contemplate a number of things as selected from different classes they still come under the *one highest* class of *things* and this is what the number has reference to when no other specification is given. [C59.7]

Part C

"The Philosophy of Logic" – A Sequel to "The Laws of Thought"

Chapter VIII

[Preface]

————————————[Probably 1857]————————————

The present work is intended as a sequel to a former publication of the author's entitled An investigation of the laws of thought upon which are founded the Mathematical Theories of Logic and Probabilities.* Upon the nature of the relation in which it stands to that work and upon the object which it is designed to accomplish it may be desirable to say a few words.

In every developed science two objects, both of which are of indispensable necessity, may be distinguished viz.: 1st the investigation of its ultimate irresolvable elements and of its primary laws 2ndly the actual construction of the science upon the basis which that investigation furnishes.

In the science of Logic the immediate subject of analysis is a certain series of intellectual operations not absolutely simple in character. It is requisite that we should determine what are the ultimate elementary operations out of which all complex operations are formed and what are the laws [to] which those elementary operations are subject. The investigation occupies the first place in the treatise on the Laws of Thought. Among the various criticisms which have been passed upon that work I have not met with any attempt to prove that the results of this investigation are erroneous. [W3.1]

But in the construction of Logic as a Science and as an Art upon the basis thus obtained (as a science in that it consists of general theories developed out of the primary laws above described an Art in that it applies those theorems to practice) algebraical forms and processes are employed. To many this has been a source of difficulty. Their employment is not however arbitrary. It rests upon the ground of an analogy proved to exist between the operations of thought in Logic and its operation within a particular sphere of

---

* Mr. M. thinks that this makes it a little too dependent on the former work.[1]

the science of number. This analogy consists neither in likeness of the operations themselves nor in likeness of the subjects upon which they are performed, but as every strict and proper analogy does – in *likeness of relations*. Agreement in formal laws is the only ground upon which any connection of method between Logic and the science of number is possible. And this agreement does not as a matter of fact exist between Logic and the science of number in general but only between Logic and a very peculiar branch of the science of number – an Algebra of the most special character yet in whose theorems the highest generalizations of Logic are embodied.

Here a question of great interest presents itself. Is the analogy which has been referred to above as connecting the intellectual operations in the two distinct spheres of Logic and of the special Algebra under consideration essential to the full development of the former science? I have certainly never regarded it as such, freely as I have employed it for the discovery of *methods*. And yet throughout the whole domain of truth natural and moral how often does light stream from one department upon another. How little is there of real isolation and independence. Assuming however that [W 3.2] this analogy is not essential to the object in view an evident consequence must follow. It must be possible to interpret within the purely logical sphere and by purely logical ideas and conceptions all the processes methods and results to which that analogy has led. Now such an interpretation is the object of the present work. I seek to bring into light and prominence the philosophical elements which in my former exposition were too much hidden beneath the veil of a symbolical notation. I think that the conclusion which these supplementary inquiries will tend to establish will be that the science of Logic even in its strictest acceptation is not founded upon any single truth but upon a harmonious and wonderfully connected system of truths. I think it will also appear that the different metaphysical principles which have been invested by different schools of logicians with distinct and exclusive predominance present themselves in that system under a new aspect and relation some of them among its primary truths but possessing only a co-ordinate authority others among its ulterior and derived consequences.

In the opening chapter of this work I have endeavoured to give a brief sketch or review of some of the most important questions which at present divide logicians and of the opinions which have prevailed respecting them. Such a review appears to me to be an almost necessary preliminary to the inquiries which will follow. If in its execution something of a controversial aspect should present itself I trust that this will not be deemed its most prominent feature.

I wrote the former book for mathematicians. The subject is of wide interest. This is intended for the general public. Mathematics will not appear except in the notes. [W 3.3][2]

Chapter IX

Table of Contents

—————————————————[Later than 1854]—————————————————

Chapter I: Of the ordinary Logic

Analysis of the Proposition (terms, copula). Conception. Terms thought of without limitation *the whole* or under limitation *part*. Principle of contradiction. Conversion of propositions. Reasoning, the dictum, the supplementary dicta of Lambert etc. The question as to the ultimate nature of the proposition whether an equation or not – also as to the ultimate principles of reasoning – whether it is the dictum or the completed dicta or the principles of identity contradiction excluded middle Sufficient reason. Design of the Laws of Thought. Contemporaneously with the development of the ordinary Logic there has been a development of mathematical science. This is reasoning exercised upon a particular subject – and the inquiry arose whether the analysis of its methods may not suggest methods more complete than those in existence before – Design of the present work, to pursue this investigation into its philosophical consequences, i.e. to inquire what light the processes of the laws of thought throw upon the philosophy of Logic – what interpretation they admit within the sphere of pure Logic.

Proposition the expression of a judgment connecting the *terms* by the *copula.**

These two laws govern all the moods of Categorical Syllogism, including under them as subordinate rules the "dictum de omni et nullo" – as well as the distinct axioms, which have been framed by different logicians as rules of the second and [C 24.1] third figures. Ib. p. 206.

—————————————

* The principle of identity is immediately applicable to affirmative moods in any figure and the Principle of Contradiction to negatives. Mansel p. 205-6.[1]

Chapter II: Of the operations of thought in relation to the science of number

Direct operations relating to number in general $x+y = y+x$, $xy = yx$, $x(y+z) = xy+xz$.

Inverse operations their nature – subtraction – division. Foundation of their laws.

Operations relating to particular number. General principle. The limitation of the subject increases the number of the laws. Laws of 0, 1. Methods founded thereon.

Laws relating to equations.

Elimination.

Chapter III: Of the Laws of Thought in Logic

Laws of Conception generally the same as of a system involving the symbols 0, 1, – of Universe 1, of Nothing 0.

Regarding propositions as equations the laws of *direct* inference are the – [sic] and therefore of inverse. Examples from laws of Thought.

Chapter IV: Interpretation of Methods

Distinction of Synthesis and Analysis both exemplified [C 24.2] above. Laws of Conceptions include the principle of contradiction with others – e.g. the indifference of order of certain operations etc. Laws of Judgment and of reasoning best considered together. – Those which belong to synthesis not readily expressible in common language – and not peculiarly interesting from their appearance as mere truisms. Those which relate to analysis [are] more important. Analysis first presents the subject under the form of a necessary proposition whose form is determined by the laws of thought by the principles of contradiction and excluded middle. These are properly the laws of *a priori* judgments independently of the nature of the premises. Secondly it shews that in the limitation of this proposition so as to produce the conclusion from given premises no more of the essence of the conception involved in these premises is employed than is involved in the primary conceptions of existence and non-existence. These are the parent[2] categories. From these are derived the categories expressed by 1, 0, $\frac{0}{0}$, $\frac{1}{0}$.

Thus a conclusion is an *a priori* judgment expressed by the laws of contradiction and excluded middle and then limited by the four categories – the *process* of limitation depending upon our contemplating each element of the original premises under the two prime categories of existence and non-existence. [C 24.3]

Chapter X

The Philosophy of Reasoning

I have undertaken the present work in the hope of supplying a defect felt by many who have engaged in the study of that peculiar development of the Science of Logic which is contained in my treatise on the Laws of Thought.

Persons unacquainted or but partially acquainted with mathematics have complained of the difficulty which they experience in endeavouring to form any intelligible notion of the object of the treatise and of the nature of its methods. Mathematicians while admitting without any exception that I am aware of the validity of its processes have expressed a not unreasonable desire to know something more of their philosophy. I have felt this desire myself and the present work is the product of those researches in which it leads me to engage. Possibly some may think that this is equivalent to a confession of error or of change of view. The former it is not. The latter it only is in so far as growth and development imply change. There are few instances if any in which the philosophy of a science has been fully matured by abstract thinking *a priori* and its laws and processes and special conclusions afterwards deduced in the order of a descending sequence. Nor perhaps are the examples many in which the converse order of ascent has been rigidly maintained from particular facts to general laws and from general laws to that higher philosophy which assigns to the different sciences their relative places and pronounces upon their nature and estimates their validity. Far more usually the methods of induction and deduction of experiment and hypothesis are mingled together or rather succeed each other in a species of continued alternation, fact correcting theory and theory guiding those inductions from fact from which it is itself to receive enlargement and accession. Such at least was the course of those investigations by which the system of Logic presented in the Laws of

Thought attained its actual form. And it is only a continuation of the same procedure to endeavour to arrive at a point from which the philosophy of the system shall appear at once more completely and more simply to view. [A91.1]

In one important respect this work will differ from its predecessor. It will be free from mathematical symbols a few illustrative notes excepted. For the peculiar objects which it contemplates the language of symbols is scarcely if at all needed. There are it is conceived persons, and private correspondence has acquainted me with some, who are interested to know all that can be known of the intellectual constitution and who yet may be unwilling or unable to pursue trains of reasoning conducted by symbols in which the laws of that constitution are if we may use such an expression embodied. Skill in the processes of art or science applied is one thing. A thorough comprehension of the grounds upon which those processes rest is another.

Among the questions of speculative interest which belong to these enquiries, the following occupy an important place. What is the nature of that constitution by which men reason? What are its members, parts and faculties? Is language an essential element? Does reasoning consist in an application of *a priori* truths or is it a process conducted in obedience to laws? If the latter, what are those laws how do we arrive at the knowledge of them what is the nature of their dominion [and] in what relation do they stand to the aforesaid *a priori* or necessary truths? Is mathematical reasoning peculiar, or is it only a special application of the ordinary rules of Logic? What is the nature of induction and the grounds of its validity? Does it arrive at convictions partaking of the nature of certitude or does it still and always remain in the realm of probability? These are among the questions which would be most likely to suggest themselves to the student some of them at the very threshold of the investigation some at subsequent [A91.2] stages of his progress in it. And there exist many others of scarcely less weight and moment to which it is not necessary here to direct attention.

The difficulty of so presenting the subject of a work like the present to the notice of the reader that it shall not deter by its novelty and technicality nor disappoint by a shallow and imperfect treatment of the subject is not slight. The mode in which I propose to meet it is the following.

I shall endeavour to present in a plain and simple exposition those first elements of the Science of Logic which consist in the definition of its subject matter, the explanation of its terms and the statement of its more obvious principles. And this portion of the work I shall aim so to write that it may serve as a brief and popular exposition of the science of Logic. I shall then go on to the enquiries which formed the more special business of this treatise and of which the design and object have been already explained. It might seem so much has been written on what are termed the elements of Logic that the former portion of this task was superfluous. Yet no treatise that has yet appeared has presented those elements in a manner accordant with the views which I have been led to form of the philosophy of the science of Logic

– none therefore in agreement with what I conceive to be the truth. I will point out only one or two particulars in which the existing treatment of the elements seems to me erroneous or defective.

Logicians recognise the tripartite division of the intellectual powers so far as their examination falls under the province of Logic into Conception, Judgment and Reasoning. And they admit in words at least that the exercise of the faculty [A91.3] of Judgment involves that of Conception and the exercise of Reasoning involves that of Judgment and therefore that of Conception also. It would seem then that the analysis of the faculty of Conception must be essential to the understanding of the operations of the higher faculties of Judgment and Reasoning. And yet this analysis is almost entirely omitted in the ordinary treatise on Logic. Sir W. Hamilton in a review of Whately's Logic[1] first pointed out this defect.

And recent treatises as compared with their predecessors certainly indicate a growing attention to the subject. But I do not know where to point to even a complete statement of the elementary operations involved in Conception still less to an analysis of the laws of that faculty. An enquiry into the nature of reasoning involves an enquiry into the nature of the processes of Conception and of Judgment but it by no means follows that the latter enquiry is only interesting and valuable from its relation to the former. [A91.4]

It is undoubtedly true that a strong distaste for all studies which are thought to be of a metaphysical nature characterizes many of the most energetic minds amongst us. But I cannot think that this is due to any deliberate and settled conviction that the human mind, laying aside all question of material profit, is not as worthy an object of speculation as the material universe. I think it is rather to be attributed to a doubt whether we possess any powers of introspection at all corresponding to those by which we look abroad, whether the subject is within our grasp, whether a veil, dark and impenetrable, does not shroud from our vision all that we should most desire to know, and leave us only a possession of truisms.[2]

The mysterious realms of thought if they are not wholly shrouded from view is only so far revealed as to become the object of controversies without profit and without [end].

An inherent vitality must belong to a study which against the force of such presumptions still commands even if only in the hearts of a few a genuine interest. The source of that vitality is I think to be found in the human relation. [A91.5]

Chapter XI

Logic

—————————————[Later than 1855, maybe 1860]———————————

There is a more general and there is a less general sense in which the term Logic is employed.

The more general definition of the term is implied in its derivation. As the word λόγος[1] signifies not only the inward thought but also its outward form or manifestation so by the term Logic in its primary and most general sense we understand the Philosophy of the Laws of Thought as expressed.

Now the expression of Thought here implied is the office of signs or symbols of which the words of common speech are the most familiar examples. And of all systems of signs this indeed is the most important. Still it has been found a matter not merely of convenience but of necessity to employ other systems of signs in particular departments of thought. Number, magnitude and their relations, the so-called affections of space the ultimate forces and elements of the material universe so far as they are at present known to us, have been represented and the thought of which they are the objects expressed by signs.

In this its highest conception therefore Logic might be said to be the Philosophy of *all* thought which is expressible by signs whatever the object of that thought, whatever the nature of those signs may be. Nor is this conception either vague or unreal. There is a philosophy of signs which governs and explains all their particular uses and applications, – which is equally manifested in the forms of ordinary speech and in the symbolical language of mathematics. The perfect idea of Logic is not that of a mere system of rules but of a philosophy from which as from a common stem all sciences whose method is deductive are developed and with which they all stand in vital connection.

But though such a philosophy exists and though it is important that it should be recognized not only as existing but as constituting the ultimate aim of all inquiry into the constitution of our intellectual faculties, it is not this which constitutes the meaning of the term Logic according to ordinary usage. For [this] purpose a *far* less comprehensive definition must be employed. Logic according to this definition is the science of the Laws of Thought as expressed in the use of the general terms of Language. Let us endeavour to explain this view.

In contemplating a group of objects we are perhaps impressed with the fact of their *likeness* to each other. We notice the several qualities in which that likeness consists, we combine them mentally in some general conception, we express that conception by a name. The things which that name represents separated in thought from all other things in the Universe constitute a *class*. Perhaps we compare this class with other classes, the notions of which have been formed by a similar process of thought. We thus become conscious of class relations. We see that one class is contained in another as a part in a whole, – a species in a genus. Hence general propositions by which such relations are expressed. Hence [C 57.1] reasoning by which from propositions thus formed, other propositions are deduced as *conclusions*. As the ground of all this procedure is the possibility of our forming the conception of Class the Logic which determines the forms and laws of such procedure, may be said to be the scientific development of the notion of a *class*.

Now a very important question here presents itself. It is maintained by some and perhaps by the majority of the professed logicians that the Logic of Class implicitly contains all Logic whatever. All reasoning for instance it is said is ultimately reducible to that act of the mind by which we apply a general truth to particular instances. According to this view the perception of the relation of genus and species forms the very foundation of reasoning. Others depreciate the importance of this relation either by denying that it is essential in any case or by affirming that it is only one of many relations which are concerned in the different processes of inference.

To illustrate these different views let us take the following example from elementary geometry viz.:

> The line A is equal to the line C
> and the line B is equal to the line C
> therefore the lines A and B are equal.

Logicians of the Aristotelian school contend that the completed form of this reasoning would be

> Things equal to the same thing are equal to each other.
> But the lines A and B are things equal to the same thing.
> Therefore A and B are equal to each other.

And the reasoning itself is said to consist in the particular application of a general truth. On the other side it is maintained that from the perceived equality of A and B to C the inference of the mutual equality of A and B is drawn directly without ascending to any more general proposition whatever. It is said that the axiom

Things equal to the same thing are equal to each other

adds nothing to the reasoning inasmuch as we could not perceive the truth involved in the general axiom unless we could perceive it in the particular instance. And some have hence proceeded to contend that all reasoning is hence of particulars.

Without entering further in this place into the different arguments which have been brought forward on the subject, I will briefly state my own views.

I think indeed that the Conception of Class is antecedent in the order of thought to all other scientific conceptions.

For instance it is prior to that conception of Number which is the foundation of the Science of Algebra. We cannot represent things under the conception of Number without first supposing them to possess that degree of likeness which permits us to regard them as units capable of repetition. But while this conception of likeness is thus involved in the very genesis of other scientific conceptions it appears to me that there exist other elements also upon which the procedure of thought may in a more direct manner depend. For instance in the example above given even when placed in the syllogistic form we see that the conception of *equality* has quite as much to do with the *reasoning* as the conception of the relation of [C57.2] genus and species or of general and particular. In the very statement of the axiom which forms the major premiss the mind proceeds according to a line of *rational* suggestion derived from the conception of equality. For to pronounce as a *necessary truth* that Things equal to the same thing are equal to each other is really to say Things equal to the same thing are *therefore* equal to each other.

In fact [in] the full and unelliptical statement of the reasoning as given above there are in reality two distinct lines of thought, one founded upon the conception of equality and exemplified in the major premiss the other founded upon the idea of the relation of genus and species and applying the conclusion established in that premiss to the particular case under consideration.

It seems to me therefore that the opinion that the mind can proceed along other lines of suggestion than that by which we descend from the general to the particular is not at all invalidated by the fact if such it be that we can throw every demonstration into a syllogistic form because in so doing we may only be throwing back into a series of major premises the elementary steps of suggestion and inference upon which the really essential part of the reasoning depends.

But while we contend for the truth that there are other conceptions and other lines of mental suggestion than those which are taken account of in the Logic of Class, it is not to be forgotten that the conception of class really occupies a position of priority in the intellectual order. That we should be able to throw all reasoning or at least a very large portion of it into syllogistic forms is at least a very remarkable circumstance. These considerations give to the Logic of Class in itself a high importance. Though it is only a particular Logic it is yet the most important of all the forms in which that higher and more comprehensive Logic, the idea of which has already been explained (article I)[2] admits of being developed.

And now we are able to explain with some degree of distinctness the object of this work. It is to develop the narrower Logic of Class but with constant reference to that higher Logic above referred to, the philosophy of thought as expressed by signs, to which it is subordinate.

If it should seem to any one that the preferable course would have been to investigate first and independently the principles of that higher Logic it may be observed that it is from its very nature incapable of being developed *a priori*. For the most part we are only able to arrive at a knowledge of the universal by means of that of the individual and the particular. By studying particular manifestations of thought we ascend to its general laws and this we do not so much by comparing particular forms and instances and selecting the truth which is common to them all as by some deeper faculty of insight enabling us when contemplating some general truth manifested under particular forms or conditions to perceive how far such forms or conditions are necessary and how far they are accidental. Thus the study of the intellectual procedure in a particular province of Thought *may* lead us to some degree of acquaintance with its *general* philosophy. And it is only by the study of these special and external manifestations that it is possible for us to arrive at any adequate knowledge of that Philosophy in itself. [C 57.3]

Again though it may not be asserted that the entire philosophy of the use of signs can be illustrated in connection with that form of the Science of Logic which consists in the development of the notion of *Class*, it may be affirmed that there is no other particular science which would answer the purpose of illustration equally well. No other science brings us so directly face to face with some of the deepest questions which are involved in the connection between Thought and Language.

With a view to the succeeding inquiries we proceed in the next place to offer some observations upon the nature and office of signs.

Nature and Office of Signs

1st. In the foregoing sections signs have been described [C 57.3a] as serving for the *representation* of things and for the *expression* of thought.

There is a more strict propriety in this language than at first sight appears. For 1st. Language is thought *uttered*. In giving a name to any class of things we first distinguish those things in thought from all others, and the distinction present in our thought when the name is given is that which we seek to fix in the name itself. 2ndly. Thought, taking its rise in those impressions which, through the constitution of our perceiving faculties, external things produce upon us, advances by the operation of our other faculties of comparison and abstraction to these general conceptions of which signs are the immediate utterance. Here the order of procedure is manifest. Things first presented to us in perception are in a certain sense reproduced and presented to us a second time in the substituted forms of language.

2ndly. It follows hence that signs serve also as *instruments* of thought.

For it has been seen that signs are representatives of things. They first *express* our conceptions of things, and then by a process of substitution founded in association stand for these things. Again they stand for things contemplated not as individuals, but as falling under the general conception of class or kind i.e. they stand for things under those relations [of] our power of conceiving which makes deductive inference possible.

Let us consider the following example of reasoning, viz.:

> All men are mortal.
> Caius is a man.
> Therefore Caius is mortal.

This argument is called a syllogism. It consists, as all syllogisms do, of: 1st. two propositions called the premises in which two terms called extremes are severally compared with another term called from its being thus the common medium of comparison, a middle term, 2ndly. a final proposition called the conclusion in which the two extremes are compared together directly. In the above example the middle term is man, the extremes Caius, mortal. Now it will be observed that the argument is valid in form independently of the *particular* meaning we attach to its terms. We must indeed understand the word *men* to represent some *class* of beings, and we must in some equally general way interpret or conceive the possibility of interpreting the other terms employed, but it is not necessary in order to pronounce the reasoning valid that we should realize in thought either the *image* of a man or the idea of mortality. The sign itself takes the place of image or idea.

But the mode in which signs serve as instruments of thought will be more fully considered in the following section.

3rdly. Signs are arbitrary as concerns their outward form, fixed as concerns their interpretation and their laws.

That signs are arbitrary as to their outward form, is evident from the diversity of languages, the same thing being represented in one language by one combination of letters or sounds and in another language by another.

That they are fixed as concerns their interpretation is a truth which is familiarly expressed in the rule that the meaning of a word or any other sign must not be ambiguous. The sign is arbitrary only in the [C 57.4] sense of being conventional. Whatever meaning is once given to it, must continue to be associated with it, if language is to be definite as a medium of communication or exact as an instrument of thought.

Again signs are fixed as concerns their *laws*. For when interpretation of a sign has been fixed, the *use* of the sign as manifested in the nature of the combinations with other signs into which it is capable of entering is fixed also. It is to be remembered that all intelligible language is organic in its structure and owes its significance not simply to the meanings of the terms employed, but to their combinations. Now it is the general rules of such combinations, – the general rules determining the variety of forms under which such combinations are *intelligible* which constitute the laws of signs. And because language is as has been said an organic structure these laws are fixed.

4thly. The laws of signs are a visible expression of the formal laws of thought.

As signs are expressions of thought the laws to which they are subject are expressions of laws of thought, and they are in a peculiar sense expressions of its *formal* laws because they determine the variety of forms which the expression of the same thought may assume. If we regard thought as consisting of certain intellectual operations, it may with the greatest propriety be said that the laws of its signs express not the conditions under which these operations are possible, but the forms which when possible their expression assumes.

For instance the conception of any collection of things as a *whole* is formed by a particular operation of thought from the conceptions of the parts which comprise that whole. From the distinct conceptions expressed by the terms animals, vegetables, minerals, we can construct that conception of a whole which is expressed by the words "animals and vegetables and minerals". The operation by which the distinct conceptions are in this way combined has its appropriate sign – the conjunction *and*. Now this operation is subject to a certain manifest law. In the conception of a whole *as such* the order in which the component parts are contemplated is indifferent. The same whole is expressed by the terms Animals, and vegetables, and minerals and by the terms Vegetables, and minerals, and animals. Hence we are led to the corresponding but derived law of the sign, *and*, viz. that the order of the terms which it connects is indifferent, and therefore that the terms themselves may be transposed. And this law determines the possible variety of forms under which the conceptions of a whole *as composed of parts* may be expressed.

In this as in other cases we observe both a law and a condition. The condition under which the operations of aggregation above described is possible, is that the parts to be conjoined be distinct, or, as we shall in future say, mutually exclusive; the law of the operation when possible is that the order of the connected terms is indifferent. While the condition relates to the possibility of the operation, the law determines the possible variety in the forms of its expression.

The intellectual operations with which Logic in the narrower sense of Art ( ) is concerned, are usually described as Conception, Judgment, and Reasoning. As to their offices it may be said [C 57.5] that by Conception we apprehend things in their different class relations, by Judgment we form propositions, by Reasoning we infer consequences. All these operations are subject to laws admitting of an outward development or manifestation in the laws of signs. As respects the collective results of this system of laws it may be said that the forms under which our conceptions, our judgments, our conclusions are expressible become determinate either absolutely or under variations which are themselves determinate.

Thus from the syllogism of Art ( ) it may be seen that there is a connection between the validity of an argument and the form of its expression. If for the sake of generality we present that argument in the generalized form:

> 1st. Premiss   All Ys are Xs
> 2nd. Premiss   All Zs are Ys
> Conclusion   Therefore All Zs are Xs

we see that the argument is valid in *form* whatever classes of things we represent by the letters Xs Ys Zs. From two premises of the forms above exemplified we can always draw a conclusion of the form also exemplified. Now this connection between the form of the conclusion and the forms of the premises in valid reasoning indicates that signs are not to be employed in an arbitrary manner, but in combinations which are so far at least definite that the nature of the connection in the premises predetermines that in the conclusion.

Methods in Logic

It has been said in the foregoing section that in virtue of the laws of thought as expressed in signs, the forms under which our conceptions, judgments, and conclusions are expressible become in a more or less absolute sense determinate. As a particular illustration of the truth, it is shewn that there is a connection between the validity of that species of argument called the syllogism [and] the form of its expression.

Now it is the object of Method in Logic to determine in a precise manner and for our practical guidance what the formal conditions of validity of inference and generally of correctness in the operations of thought are. Accordingly the methods which have been proposed in the Science of Logic e.g. the Aristotelian method, the method of Lambert and of Sir W. Hamilton and the mathematical methods of Leibnitz, Drobisch, De Morgan etc. differ: 1st. as to what they assume the primary laws of thought to be; 2ndly. as to the nature of the rules, which upon their assumed basis of law they furnish for our practical direction; 3rdly. as to the extent of those rules. Under the last head of difference would fall such questions as the following viz. whether the rules in question apply to all the processes of thought or only to that of reasoning, whether in their application to reasoning they take account of all the forms of inference or only of the syllogistic form etc.

It is necessary that some account should be given of the characteristics of these various methods in order that the precise nature of the method of this work may be understood. But it may here suffice to speak of these with more particular reference to the operations of reasoning.

The Aristotelian Logic

As an exposition of reasoning the Aristotelian Logic may be [C57.6] said to consist of a collection of valid forms of syllogism arranged in a mnemonic scheme and as concerns their ground or origin referred more or less directly to a certain fundamental principle, known in the schools as the *dictum de omni et nullo* viz.:

> Whatever is affirmed or denied universally of any class of things Ys, in which a certain other class of things Zs is included, may be affirmed or denied in like manner of the class Zs.

or in briefer terms:

> Whatever is affirmed or denied of a genus may be affirmed or denied of any included species.

As concerns this principle a little attention will perhaps shew that it contains a condensed statement of the office and relation of the different parts of a syllogism.

We might give to it the more explicit form,

> Whatever in the first premiss is affirmed or denied universally of any class of things Ys, in which by the second premiss a certain class of things Zs is included; may in the conclusion be affirmed or denied in like manner of that included class Zs.

Thus in the example of Art ( ) it is in the first premiss affirmed of All Ys that they are Xs. By the second premiss all Zs are included under the class Ys. The conclusion affirms of all Zs that they are Xs.

In the syllogism,

$$\begin{array}{l} \text{No Ys are Xs} \\ \text{All Zs are Ys} \end{array}$$

Therefore   No Zs are Xs

what is denied in the first premiss of the genus Ys, in which by the second premiss the species Zs is included, is in the conclusion denied of the species Zs.

There are forms of syllogism to which the *dictum de omni et nullo* does not directly apply.

Logicians take account of three and some contend for four distinct forms, technically called "figures" of Syllogism, dependent upon the relative positions of the terms. In the scheme which recognizes four figures these are distinguished as follows. Designating that premiss which contains the predicate of the conclusion as the major, and that which contains the subject of the conclusion as the minor, then,

In syllogisms of the first figure the middle term is the subject of the major, and the predicate of the minor premiss. This is exemplified in Art. ( )

In syllogisms of the second figure the middle term is made the predicate of both premises. Example:

$$\begin{array}{l} \text{All Xs are Ys.} \\ \text{Some Zs are not Ys.} \end{array}$$

Concl[usion]:   Some Ys are not Xs.

In syllogisms of the third figure, the middle term is made the subject of both premises. Example:

$$\begin{array}{l} \text{All Ys are Xs.} \\ \text{All Ys are Zs.} \end{array}$$

Concl[usion]:   Some Zs are Xs.

In syllogisms of the fourth figure, the middle term is the predicate of the major premiss and the subject of the minor premiss. Example:

$$\begin{array}{l} \text{All Xs are Ys.} \\ \text{All Ys are Zs.} \end{array}$$

Conclusion: Some Zs are Xs.

Now it is only to syllogisms of the first figure that the dictum applies directly. To reduce the other figures of syllogism under its dominion it is necessary either to *convert* one of [C57.7] the premises i.e. to alter its form in some *legitimate* way, but so that the subject and predicate shall change place, or to call in the aid of some other principle such as the reductio ad absurdum.

For instance if in the above example of the syllogism of the third figure we convert the minor premiss into the implied though less general proposition, Some Zs are Ys we have by the application of the *dictum*:

All Ys are Xs.
Some Zs are Ys.
Therefore   Some Zs are Xs.

In the same way if in the example of the syllogism of the second figure we convert the major premiss into the implied and equivalent proposition All things which are not Ys are not Xs, we have by the direct application of the *dictum*:

All things which are not Ys are not Xs.
Some Zs are not Ys.
Therefore   Some Zs are not Xs.

Or by a reductio ad absurdum in the following manner. If the conclusion Some Zs are not Xs be false its contradictory All Zs are Xs must be true. Substitute this for the minor premiss and we have by the application of this dictum:

All Xs are Ys.
All Zs are Xs.
Therefore   All Zs are Ys.

But this contradicts the *given* minor premiss. Whence the supposition that the conclusion Some Zs are not Xs is false leading to a contradiction of one of the data it follows that that conclusion is true.

Beside the *dictum de omni et nullo* the Aristotelian theory recognizes also a number of derived conditions of validity in the syllogism e.g. that from negative premises nothing can be inferred and it supplies, though this is a later growth, a mnemonic classification of the valid forms of syllogism designed to enable the student to dispense with the application of the *dictum* or of any other principle.

We have remarked that the *dictum de omni et nullo* is directly applicable to syllogisms of the first figure only. Lambert and others have endeavoured to supply the defect by the invention of supplementary *dicta* applicable directly to the other figures so as to render any process of reduction unnecessary. Thus syllogisms of the second figure may be considered as expressions of the principle Two classes of things (subjects) which differ as to the possession of a certain mark or quality (predicate) are distinct.

All such dicta it is obvious agree in being like that of Aristotle only condensed forms of syllogism.

The merits and the defects of the Aristotelian theory are both due to the same cause. We must regard it less as a Science than as an Art – and to a great degree as a mnemonic Art. It is rather a Natural History of the forms in which human thought has been actually developed than an inquiry into the possible – the universal – forms in which thought admits of being developed in virtue of the constitution of the human mind. [C57.8] But this very limitation gives to it within the proper range of its application a character of directness which no mere perfect system can possess. We might perhaps adopt a mechanical analogy and say that in the ages of its early culture human thought wore for itself certain tracks or grooves in which though not constrained by law it became dispersed by habit and association to run, and that the genius of Aristotle first sought to map out those tracks and to determine their plan. The modern extensions of the theory of the syllogism gain their generality by the introduction of forms which though legitimate in themselves have none of the sanction of ancient use.

Sir W. Hamilton's Theory of Syllogism

It is the express postulate of Sir W. Hamilton, as it is the implied postulate of every professed extension of the theory of syllogism, that we be permitted to state explicitly what is thought implicitly. Logic, says Sir W. Hamilton, must be an unexclusive reflex of thought (Discussions p. 640*).[3]

Our concern, it is thus affirmed, is not merely with the forms which syllogism has tended to assume through the actual structure of language or in which it has been fixed by the authority of Aristotle and of the schools, but with all the forms which it can assume as an expression of the mental act which consists in deducing a relation between two conceptions (extremes) from two given relations in which they stand to a third conception (middle term).

For instance in that act of Judgment which is expressed by the proposition

>   All Xs are Ys

we contemplate the class Xs as *included* in the class Ys. But [it] is clearly a *possible* act of Judgment to contemplate two classes as co-extensive and so to construct in thought the proposition

>   All Xs are All Ys.

We in effect do this when we realize by a single act of thought the identity of equilateral and equiangular triangles.

Now the Aristotelian Logic does not admit in syllogism any such proposition as the last.

In affirmative propositions while it permits the subject to be either universal or particular it always supposes the predicate term particular, its two forms of affirmative propositions being

> All Xs are (some) Ys.
> Some Xs are (some) Ys.

Here then, Sir W. Hamilton steps in and claims that the same extension be permitted to the predicate term as to the subject. To the two last forms of affirmative propositions he would add the following:

> All Xs are All Ys.
> Some Xs are All Ys.

The principle asserted is known in recent controversial writings as the principle of the thorough quantification of the predicate!

It is a consequence of this extension to the predicate of a proposition of the same variations of degree with respect to quantity as belong to the subject that the distinction between the two terms becomes one of position merely. Accordingly a proposition becomes in the language of Sir W. Hamilton "an equation between its subject and predicate." All that is characteristic in his system may be regarded as following from this fundamental view of the import of a proposition.

In applying this view to the theory of Syllogism Sir W. Hamilton [C57.9] divides Syllogisms into two classes: 1st. those which involve in their premises a proposition both terms of which are taken universally as All Xs are All Ys, 2ndly. Those of which the premises consist of propositions each of which is particular in one of its terms at least. For each of these classes of propositions he assigns a distinct canon of inference. To the first class he assigns the canon. Inasfar as two notions either both agree or the one agreeing the other does not, with a common third notion; in so far these notions do or do not agree with each other. To the second class of Syllogisms he assigns the Canon. "What worst relation of subject and predicate subsists between either of two terms and a common third term with which one at least is positively related; that relation subsists between the two terms themselves." In this scheme of comparison of relation as explained under the sanction of Sir W. Hamilton by his pupil Mr. Spencer Baynes, a negative quality is a worse relation than a positive, and a particular quantity a worse relation than a universal.*

Sir W. Hamilton represents the acknowledged laws of the several figures of syllogism, such as the *dictum de omni et nullo* for syllogisms of the first figure as evolved out of the second of the above general canons.

---

* New analytic of logical forms, p. 74.[4]

He has further devised a peculiar notation for the expression of Syllogisms and more especially for the purpose of presenting the procedure of thought indifferently in either of two forms which in common language might from the nature of their expression be termed the concrete and the abstract, but which Logicians term Extension and Intension. In the mode of extension it might be said that the class "men" form a part of the class "mortal beings". In the mode of intension it might be said that "mortality is an attribute of humanity", and it is seen that the order of the terms is inverted. Now any system of notation which expresses terms by letters and at the same time explicitly assigns to those terms their implicit *quantity* enables us to read propositions either backward or forward and therefore by substituting the abstract for the concrete and making a corresponding change in the meaning of the copula to read it in intension as well as in extension.

The Theory of Professor De Morgan

Professor De Morgan accepts with Sir W. Hamilton the principle that Logic is the inclusive reflex of thought but he differs from him as to the nature of the elementary propositions which constitute syllogism – and hence as to the theory of its form.

In his system Sir W. Hamilton's first species of Syllogisms characterized by the presence of at least one proposition with both its terms universal finds no place. Regarding propositions of this class as compound he resolves All Xs are All Ys into the two component propositions

All Xs are Ys
All Ys are Xs,

either of which singly is admissible in syllogism. Thus far the system of Professor De Morgan may be said to differ from that of Sir W. Hamilton by way of *defect*. [C 57.10]

But it also differs by way of excess. Professor De Morgan admits and Sir W. Hamilton does not admit what Logicians term "privative conceptions" as the terms of syllogism proper.

To explain this distinction it must be observed that from the conception of any class of things constituted as such by the possession of a given common attribute, we can by a possible act of thought pass to the conception of the class which is characterized by the absence of that property. Thus from the conception of things animate we pass to that of things inanimate. The relation between these conceptions is expressed by saying that the one is positive the other privative, the general formula being Xs positive – not-Xs privative. It may be added that this operation has always reference to some implied sphere, or to use the language of Professor De Morgan, Universe of thought. In Chemistry we distinguish the metallic and the non-metallic, the Universe of thought being "elementary substances." When the universe of thought is not thus limited, it must be understood to comprehend all existence.

Professor De Morgan thus admits in syllogism such premises of the forms

> All not-Xs are Ys
> Some not-Xs are Ys
> No not-Xs are Ys
> Some not-Xs are not Ys

Premises in which Xs are the direct subject of thought. It need scarcely be remarked that if in both premises not Xs are spoken of we might regard that conception as positive, or rather the destructive of positive and privative would not necessarily arise. But Professor De Morgan, in admitting Xs and not-Xs together, admits conceptions which are *essentially* privative.

Professor De Morgan also gives a symbolical expression to syllogism which differs from that of Sir W. Hamilton in one very important respect. It is so contrived that by cancelling the symbols which belong to the middle term and applying a certain rule of interpretation that relation between the extremes which constitutes the conclusion [is] presented. It is therefore designed to be and it is an instrument as well as a form of expression. But this advantage is purchased at the grave sacrifice of a want of uniformity in the interpretation of the symbols, and I believe it will in examination be found that its essence as a method consists not in its symbols *as signs of thought*, but in its accompanying rules, viz. the rule determining to what premises it may be applied, and the rule of interpretation referred to above.

Mr. Morgan adopts the following notation.[5]

1st. Let x mean not-X.

2nd. Let X totally spoken of be X) or (X, partially spoken of be)X or X([6]

3rdly. Let a negative proposition be denoted by one dot an affirmative by two or none.[7]

Thus

All Xs are Ys	would be expressed	by X))Y or y))x
No Xs are Ys		by X)·(Y
All not-Xs are Ys		by x))Y
Some not-Xs are Ys		by x()Y

and so on.[8]

In syllogism it is assumed that the premises are not both particular and that the terms of particular premises are both universal. The premises are then connected together by placing the middle term between the extremes and giving it on each side that mark of quantity by which it is connected with the corresponding extreme. [C57.11]

Thus the premises

> All Ys are Xs
> All Zs are Ys

are expressed in the form

> X((Y((Z

if we cancel the middle term and its marks of quantity we have X((Z, i.e. All Zs are Xs, a cancel conclusion.[9]

Now Mr. De Morgan makes this rule general by introducing the conventions that X)(Y shall represent, not as in accordance with the previous notation we should suppose All Xs are all Ys, but some things are neither Xs nor Ys that X(·)Y shall represent, not as we shall suppose some Xs are not some Ys but everything is either some X or some Y.[10]

Thus the symbols are interpreted by a rule determining the dependence of the extremes in the conclusion upon the extremes in the premises, not the rule deduced from the interpretation of the symbols.[11]

It will be observed that the differences which present themselves in comparing the Aristotelian theory of syllogisms with the two above noticed arise in part from definition. Now definitions which are not very important in themselves may become important from their connection. Thus under the hypothesis that syllogism is the universal type of inference any differences in our view of what syllogism is assumes a degree of apparent consequence which they could not possess under a different hypothesis. It is a doctrine of the Aristotelian logic that there is no inference from two negative premises and this has almost become proverbial in the form of negatives prove nothing. Take however the premises:

> Some Ys are not Xs
> No Zs are Ys

and though neither the Aristotelian forms nor these of Sir W. Hamilton recognize any inference as possible, we may; there is a conclusion legitimate in thought viz.:

> Some things which are not Zs are not Xs

a conclusion however involving in its expression a privative term, we find here the error of a too narrow definition.[12]

It may be said that not only a Logic which shuts out such conclusions cannot be the unexclusive reflex of thought but that in so far as it is supposed to be such it must lead to a contracted and therefore erroneous view of the functions and the province of reason itself.

Observations of a similar character might be made upon some portions of the system of Professor De Morgan, upon his exclusions of propositions where terms are both universal upon his limitation of the import of names.

All restrictions of this kind are a practical contradiction of the postulate that we are permitted to state explicitly what we think implicitly for in the syllogism of pure thought we can make direct use of propositions which are universal in both their terms, we can employ names without any other restriction than that they are names, we can and do make use of privative conceptions. If it be said that it is only under such restrictions that we can reduce the conditions of inference to general [rules], like those of Sir W. Hamilton or symbolical rules of the same nature as that of Professor De Morgan, it must be replied that this very need of restrictions not founded in the nature and the possibilities of thought is a sufficient proof that the methods which [C57.12] have been employed are defective in their very foundation. Their differences are such as to throw doubt upon the very question. Is there any universal type of inference?

Of the Ultimate Laws of Thought

The name of Sir W. Hamilton is connected with other and perhaps more important views on the science of Logic than those which have been detailed above. His earlier writings on the subject contain an explicit statement of views which those who in recent years have sought rather to deepen the foundations than to enlarge the boundaries of Logic have accepted.

These views may be thus stated:

1st. That it is neither solely nor principally the object of Logic to investigate the Laws of Reasoning, that it is concerned and equally concerned with thought in all its processes.

2ndly. That the ultimate laws of thought are the three following, viz.—1st. The principle of identity as expressed by the formula Every X is X.—2ndly. The principle of contradiction, viz. It is impossible to contemplate any object of thought as possessing and not possessing one attribute at the same time. Its formula is No Y is a not-Y.—3rdly. The principle of excluded middle, viz. Of every object of thought we may predicate either any proposed attribute or its opposite, there being no *middle* course. Its formula is Every X is either Y or not-Y.

And these it is affirmed are the ultimate laws of all thought. Let us for a moment consider their real nature.

It is evident that they have no worth as expression of material truth, for it adds nothing to our knowledge of things to be informed that every man is a man and that every tree is a pine or not a pine. Their real office is to determine the forms of necessary propositions i.e. the forms of propositions which are true because of their form, not because of the nature of the material conceptions which they connect. And it is important to observe that they do this however many or few be the conceptions which they connect. For instance if we form the three conceptions expressed by the words wood hard white we may by the formula of excluded middle construct the proposition.

> Wood is either hard or white,
> or hard and not white,
> or white and not hard,
> or neither hard nor white.

In doing this we might begin with the proposition

> Wood is either hard or not hard

and then resolve by the same formula the conception of things hard into those of things hard that are white and things hard that are not white, and the conception of things not hard in the same way.

But though the above principles determine the forms of necessary propositions it has never yet been shewn how they determine the forms of dependent propositions such as all propositions expressive of logical inference are. The conclusion of a syllogism is true, if true at all, in consequence of the assumed truth of the premises, not in virtue of its mere form. [C57.13]

Whether the *dictum de omni et nullo* is the true, the only principle of syllogistic reasoning or not, it at least appears that it is a principle of reasoning, that it contains an element of formal truth different in *kind* from the elements which are contained in the principles of identity, contradiction and excluded middle. The logician who maintains that all *reasoning* is an application of the former, and they who hold that all *thought* including reasoning, is governed by the latter differ irreconcilably. And yet there are some who profess to hold both doctrines.

In the principles of identity, contradiction and excluded middle, some are of opinion that another principle ought to be added, the principle that nothing exists in thought as in the world without a determining reason – that every conclusion must rest on sufficient grounds. But this is a material not a formal principle and has nothing whatever to do with the question, what are the forms of propositions? What are the rules of inference?

Perhaps this survey may be thought to warrant the conclusion that a theory of the intellectual processes might [have] to recognize laws of reasoning as well as laws of conception, and of judgment – that it should equally determine the forms of propositions which are necessary in themselves, and the forms of propositions which are necessary as consequences of propositions which have gone before.

The object of all science is indeed the search of unity – but it is the unity which consists not in giving to any one portion of truth a predominance to which it has no claim not in making different portions of truth appear less different than they are, but in that connection which truths really different may possess, *as parts of a system*. The unity of science is an organic unity.

Of the Method of This Work

The unprogressive character of this science of Logic has been its most frequent and most just reproach. And the question naturally arises is this character fixed in the nature of the science itself or does it arise from its method?

Perhaps there might be some ground for saying that it is fixed in the making of the science itself – that the relations of class *as such*, the laws which govern the use of general names independently of the particular meaning of those names constitute too narrow a field of enquiry to permit of such being added to that knowledge of them which we owe to Aristotle.

But the history of another science which equally deals with abstract conceptions expressed by science must throw no doubt upon such a conclusion. The Science of Number is not occupied about conceptions which would be generally considered wider or more universal than those of Logic. And yet it has been and is to this day an eminently progressive branch of human knowledge. It has in its wondrous course made known to us the laws of the Universe, it is in its actual state the greatest of all the intellectual monuments of time.

Now we propose here to enquire what the method of this science of Number has been, and hence so far as the light of analogy may serve for our guidance to determine what the methods of that other science which is occupied about the [C 57.14] relations of Class ought to be. And the ground upon which we propose to do this is that *both* sciences are subordinate to that higher Logic which has been defined as the philosophy of the Laws of Thought as expressed by science; that either may therefore serve to illustrate its principles.

It is important that this should be understood. Logicians have with something like indignation protested against the supposition that *any* light can be thrown upon the principles of Logic by those of Algebra. To attempt this they say, is to set aside the true relations of the sciences, to subordinate the general to the particular. Now if in the narrower sense of Art ( ) Logic be defined as the science of relation of Class, it is not true that it stands to Algebra or the general science of Number in the relation of the general to the particular, because the conceptions with which the two sciences are conversant are different in kind. But if Logic be defined in the higher sense of Art ( ) as the science of the relations of all thought expressible by the signs, then though it be indeed true that the laws of such a science are not subordinate to or dependent upon those of any particular science, yet our *knowledge* of its laws must depend upon our knowledge of the particular sciences in which these laws are manifested. The order of the relation or as it has been termed the Hierarchy of truth is one, and the order by which we arrive at the knowledge of them is another.

The Method of Algebra

The method of Algebra involves the following elements:

A. A recognition of the fundamental operations to which the conception of number is subject.

Of these, two are direct viz. addition, multiplication and two respectively inverse to these viz. subtraction, division.

There are indeed other operations formed by combinations of these or in some other way dependent upon these, but these alone are primary.

B. A recognition of the *relations* to which the conception of Number [is subject] and more particularly of the relation of *equality*.

C. A recognition of the laws which determine the equivalent forms of expression arising from the performance of the operation referred to in A, – equivalent in the sense that the results which they express fall necessarily under the relation of equality B.

And here we must distinguish between those elements of form which are arbitrary and those which are not arbitrary.

The arbitrary elements are the signs by which we express the operations themselves. We express numbers by letters, the operation of addition by +, the operation of multiplication by × or by the juxtaposition without connecting signs of the letters by which the numbers are expressed, equality by = and so on.

All this is arbitrary. Numbers might be represented by any other system of marks. The operation of addition might be represented in the same way as that of multiplication now is and *vice versa*. And this freedom is due not to any peculiarity of the science of Algebra, but to that general principle of the [C57.15] use of signs which has been studied in Art. ( ), viz. that signs are arbitrary as to their outward form.

The elements which are not arbitrary are those which do not depend upon the individual form or structure of the signs employed but upon the laws of their combination.

For instance whatever numbers are expressed by the letters x and y we have

$$x + y = y + x$$

The law of thought here expressed is the following, viz. the order in which we think of the component members which by addition make up a sum is indifferent. Now this is a law which is independent of any conventions as to the mode of expressing the component numbers themselves or the connecting operation of addition – but when the mode has been arbitrarily fixed the law of thought becomes a law of expression and determines the variety of equivalent possible forms. If we choose to express the addition of two numbers by the juxtaposition of their singles the law would assume the following expression, viz.:

$$xy = yx$$

But it is easy to see that it is the same *essential* law which is manifested in both.

D. A recognition of the *axioms* founded in the relation of equality.

There are such principles as the following viz.:

If equals be added to equals the whole are equals.
If equals be subtracted from equals the remainder is equal.If equals be multiplied by equals the products are equal.

There is an important distinction between these axiomatic laws and the formal laws referred to in C. The latter expresses the forms of what may be termed *necessary* equivalents the former of a dependent or inferred equivalence.

The equality expressed by

$$x + y = y + x$$

depends not for its truth and validity upon what particular numbers are expressed by x and y, but when from the given equations

$$x = y \qquad\qquad w = z$$

we infer the equation

$$x + w = y + z$$

we arrive at a result which is not true because x, w, y and z are numbers, but because they are numbers connected by given prior relations.

Upon this basis of the expression of numbers, of the operations to which the numbers are subject and of the relations in which they stand by signs and upon the determination of the laws of these signs the method of Algebra is founded. Now there is nothing in the *general* character of that method which in a peculiar sense restricts its application to the Science of Number. Wherever we have to do with the analysis of expressed Thought our object should be first to determine the primary intellectual operations of which it consists secondly to investigate the fundamental laws of these operations thirdly to give formal expression to both in the sign by which the Thought is expressed, and the laws by which the signs are governed.

The procedure of the particular Science of Algebra serves to illustrate these principles which in reality belong to Logic itself in the higher sense of that term explained in Art. 1. [C 57.16] and hence flows into all its particular manifestations.

The Science of Algebra involves also in some of its processes another principle distinct from those above explained but deserving particular attention.

The formal laws of science are determined from the nature of the operations which those signs represent. Those operations are conceived in thought as capable of being realised in the world of things. Thus the addition of two numbers is an abstract representation of the collection of two groups of numerable things into a single group and this is an operation which can always be conceived possible whatever the number of the things in either group may be.

Hence the formal law of the operation of addition expressed by

$$x + y = y + x$$

imposes no restriction upon the numbers represented by x and y. But the operation of subtraction denoted by the sign − does presuppose the condition that when a number x is to [be] diminished by another number y, x should be greater than y. The equation

$$x + z - y = x - y + z$$

is perfectly intelligible and necessarily true when x, y, z denote numbers of which x is greater than y and we can realize it or conceive it to be realized in the world of things in a variety of ways. But are we permitted to employ the above equation as if it were true i.e. to substitute one of the forms connected by the sign = for the other without such implied condition connecting[13] the numbers x and y?

Now the method of Algebra assumes that we may do this − that in the procedure of the intellect the formal law is independent of the conditions of restrictions imposed upon its *material* realization. All experience shews that this may be safely done once that the data of the process of reasoning within the province of Algebra are expressed by its symbols, and the formal laws become everything. The result is always found to be the same as if the conditions of *material* realization had been attended to throughout. Now it is a very important question of general Logic perhaps the deepest question of all Logic whether this procedure is founded in the very constitution of the intellect or to speak more precisely in that part of the intellectual constitution in virtue of which we reason by the use of signs.

I will at once say that I think that it is thus founded, and I do this rather upon the ground of a large induction upon the actual processes of mathematics than of that of the perception of the truth as axiomatic. But quitting this somewhat conjectural reason I briefly state to what extent I think the principle may be adopted in general Logic without assuming any position which would not be recognized *generally* as true.

It will then be admitted that it suffices in expressing processes of thought to attend only to the formal laws of its symbols provided that in so doing we never transgress the conditions under which the formal laws were themselves determined. The procedure indeed is formal but each step of it is only representative of an external reality – each step we might conceive to be realized in the world of things. The thought and reasoning of common life are of this kind. We do not always think that which our words express but it is always assumed that our words express something which is capable of being thought. [C57.17] If we employ only the association of words we do so with the understanding that such an association represents a conceivable association in the world of experience.

It is evidently possible that in the world of experience there may exist different systems of things corresponding to which there may exist in the world of thought different corresponding systems of conceptions and ideas. And considering two such systems it is possible that the formal laws may be the same while the conditions of their realization in the world of experience may be different. Let us endeavour before proceeding further to exemplify this remark.

It has been seen that when we form the conception of simple number the possible operations of thought in dealing with this conception are limited by conditions. The operation of subtracting one number from another is possible only on the supposition that the number to be subtracted is the smaller. But when we think of number not *simply* but associated with the idea of linear extension in space the condition above referred to ceases to be necessary. If we represent by $+x$ $+y$ etc. distances measured along a line in one direction from a given point and by $-x$ $-y$ distances measured from the opposite direction from the same point then $x - y$ becomes interpretable whatever may be the absolute magnitudes of x and y.

At the same time all the formal laws of signs are the same in the one system as in the other – in the system of simple number as in that of numbers affected with another attribute derived from our acquaintance with space and direction. There are indeed other affections of Number besides this one which would equally serve the same purpose of illustration but all suppose equally an extension of our powers of conception through the means of larger experience without affecting the formal laws of thinking which govern thought as exercised in the domain of that lesser experience which presents Number only in its simple and unaffected essence.

Now let us consider the above case in connection with the principle already affirmed that "it suffices in expressing processes of thought to attend to the formal laws of its symbols provided that in so doing we transgress not the conditions under which the formal laws were themselves determined." Manifestly it follows that in our reasonings about simple number we may proceed with the same freedom as if we were reasoning about number under affections which do not affect its formal laws but enlarge the conditions under which our processes founded upon these laws admit of interpretation.

And generally if any system of symbols subject to formal laws express thought under conditions of interpretation which are not essential to all thought obeying the same formal laws the conditions of interpretation do not impose any necessary restriction upon the processes which the formal laws sanction.

In all this we have but an exemplification of one great proof viz. that the intellectual procedure as governed by formal laws and the power of conception whether as limited by its own constitution or by the constitution of the world of experience are to a certain extent independent of each other. As to the nature of this independence all our actual knowledge tends to the conclusion that though the intellectual procedure is governed by formal laws for our knowledge of which we are [C 57.18] indebted to our powers of conception it is not subject to the limitations and restrictions to which the latter are subject.

As a memorable example of this we are undoubtedly able to express in symbolical forms the solutions of many of the dynamical problems of a universe not restricted by the condition that space exists in three dimensions only – and therefore to us inconceivable.

But the extent to which we design to carry this principle here is simply to the assertion that conditions of interpretation which in the world of experience are not *necessarily* associated with the particular scheme of formal laws ought not to restrict the formal procedure of thought.

One observation yet remains. There are perhaps no cases in which that freedom from dependence upon conditions of interpretation which the above principle claims is really necessary as a condition of arriving at *results* of thought. It may be regarded as certain that all legitimate consequences of reason may be obtained by procedure of thought which while governed by formal laws never transgresses any actual conditions of interpretation. The freedom claimed is therefore valuable partly for the practical convenience which it secures, partly and perhaps chiefly because it practically illustrates a great mental law.

Method of This Work

The method above described will be employed in the development of the Logic of Class in the present work with such differences as arise from the difference of the subjects, differences which do not affect the essential character of the method. It may be proper to state beforehand the principal steps of the application.

1st. We shall investigate the formal laws of the operation involved in conception.

2ndly. We shall investigate the formal laws arising from the *relation* which connects the terms of propositions.

3rdly. Upon these two sets of formal laws the procedure of the science will be founded.

And now as concerns the *mode* in which the knowledge of the formal laws above mentioned enables us to accomplish this end partly this consists in the *direct* application of the individual laws themselves – partly in the application of a certain fundamental truth or principle in Logic which the laws themselves considered not individually but collectively and as a system make known to us. On the nature of this truth and the nature of its evidence I will say a few words.

It is fully recognized that the formal procedure of reasoning and it might be added of thought generally does not depend upon the distinctive meaning of the class terms employed. An argument the conclusion of which expressed some inferred property of "red things" would be equally *valid* if the word "blue" were substituted throughout for "red" or even if for "red things" we resubstituted oxen. Of course such change might make the premises false, but it would not therefore render the argument invalid. The conclusion would still be a rigid consequence of the premises *supposed true*. Now if the correctness of reason do not depend upon the distinctive meaning [C 57.19] of such terms as "red", "blue", "men", "animal" etc. the question arises does it in any way depend upon their meaning? The study of the formal laws of conception leads to a remarkable answer to this question. It shews that the formal laws of all conceptions which admit of expression in the *general* terms of language are precisely those which are common to the conceptions of *existence* and *non-existence*. This is the general truth above referred to. The entire procedure of the Logic of Class is thus made to depend either upon the direct application of the formal laws themselves or upon reduction of all conceptions expressed by the general term of language to their fundamental elements, the conceptions of existence and non-existence, which are the ground of these laws. Of these methods the former is synthetical the latter analytical. Both will be sufficiently illustrated in future chapters of this work.

It may be noticed that the principles of identity, contradiction and excluded middle which in the view of Sir W. Hamilton are the sole fundamental laws of Thought have their formal representatives or equivalents in the laws above referred to; but of the system of those laws they constitute only a part. Not only are other laws involved in that system essentially, but the very constitution in virtue of which it is a system must be regarded as essential also.

Results of This Method

Of the ulterior conclusions of the above method I shall here notice only the one which I deem the most important viz. the answer which it gives to the old question What is the general type of inference?

Before explaining what this answer is, it will be necessary to enter into one or two preliminary details.

Every process of reasoning consists in deducing a legitimate conclusion from given premises. In the Logic of Class which alone we are here considering each premiss is a proposition connecting *terms* by means of the copula is or are and these terms are either simple or complex – in the latter case being formed by that mental process, by which different conceptions are combined together into a single conception. Thus the premises consist of expressed logical relations among conceptions and any legitimate conclusion will express a deduced relation among those conceptions or among *some* of those conceptions. In the case of the syllogism for instance we have two premises expressing relations between three conceptions two only of which are retained in the conclusion, the remaining one (middle term) having been got rid of or to speak technically eliminated. The most general idea of Logical inference here suggested is that of the process of thought enabling us to eliminate from any given system of premises however complicated and however numerous any of the conceptions they may involve and to express the whole of the logical relation connecting the remaining conceptions according to any legitimate order.

Now the formal laws of thought and chiefly those called the principles of contradiction and excluded middle enable us to form a *necessary* proposition connecting any proposed conceptions whatever in perfect independence of any relation established between them by premises. Taking for instance [C57.20] the two conceptions "men" "rational beings" we may at once say:

Men are either rational or not rational,

meaning thereby that every individual man belongs *necessarily* to one or the other of two alternative classes composing the predicate term. If we introduce another conception animal we have the necessary proposition:

(A)   Men are   either rational animals
                   or rational not animal
                   or not rational but animal
                   or not rational not animal

expressing that every individual man belongs of necessity to some one of the four alternative classes forming the predicate term. If we introduce another conception we should be able to construct a necessary proposition involving eight alternative classes in its predicate term and so on. Speaking generally we see that these necessary propositions consist of a subject term which is perfectly arbitrary expressing any conception whatever, simple or complex, a predicate term expressing the possible alternatives which can be formed from any other conceptions and a connecting term of relation referring each individual in the subject to some one or other of the alternatives involved in the predicate. We notice further that while the class which constitutes the *subject* of the proposition is, in respect of quantity spoken of universally, the several alternative classes which comprise the predicate terms, are spoken of indefinitely. While in the proposition (A) every man is referred to some one of the classes composing the predicate, it is left wholly undetermined whether any of the individuals in one of these classes e.g. the class "rational animals" is included in the class men. We shall express this by saying that the class which forms the subject is in the category of the universal and each of the alternative classes forming the predicate in the category of the indefinite.

Now the consequences to which the method of this work leads with reference to the question of the universal type of logical inference is the following, viz.:

1st. A logical conclusion is always in the form of a necessary proposition modified by means of the premises.

2nd. The nature of the modification is the following viz. the alternatives which in the predicate term of the unmodified necessary proposition are all in the one category of the *indefinite* are each determined under one of four categories viz. the universal, the non-existent, the indefinite, the impossible.

When any one of these classes is in the category of the universal it is implied that every individual contained in that class is contained in the class represented by the subject terms when in the category of the non-existent that none of its members are contained in the subject term; when in the category of the indefinite that each member may or may not be contained in the subject class; when in the category of the impossible that the class does not exist at all.

3rdly. Such determinations of the alternative classes of the predicate term under the four categories above explained are quite independent of each other. Any alternative class may be determined under any category quite irrespectively of the categories under which the other alternative classes are determined. [C 57.21]

It will be observed that, as the subject term of the conclusion and the elementary conceptions from which the alternatives in its predicate term are formed may be chosen arbitrarily, we may have a considerable diversity as to the forms of the conclusion deducible from given premises.

To illustrate this doctrine I propose to examine a few of the conclusions deducible from the premises of an ordinary syllogism in Barbara.

The given premises being

> All Ys are Xs
> All Zs are Ys

we have first the conclusion

> All Zs are Xs.

Now this is no other than the necessary proposition

> All Zs are either Xs or not Xs

modified by placing the not Xs under the category non-existent. Again we can obtain a conclusion in which the middle term is presented [as] subject, while the alternative classes formed from the two extremes and placed under proper categories constitute the predicate term. Irrespectively of the premises we should have the necessary proposition

> All Ys   are Xs which are Zs
>   or Xs which are not Zs
>   or not Xs which are Zs
>   or not Xs which are not Zs.

Now the conclusion sanctioned by the premises is found by modifying the above in the following way viz. by placing the first alternative class in the predicate under the category of the universal the second under that of the indefinite the third under that of the impossible the fourth under that of the non-existent, the conclusion may be resolved into that of the following, viz.:

1st. The Ys consist of All Xs that are Zs and an indefinite remainder (some, more or all) of Xs that are not Zs.

2ndly. There exists no class of things formed by Zs that are not Xs. It will be observed that the second of these results is equivalent to the ordinary conclusion of the syllogism.

It may perhaps be objected to the above theory that it presents the conclusion of a process of logical inference in a form which does not admit of direct expression in a single proposition of common language. In the example last given the presentation of the class of Zs that are not Xs under the category of the impossible necessitates the employment of a distinct proposition of common language for its expression: further it may be said that the form in which such proposition assumes the declaration of the non-existence of the class of Zs that are not Xs is one which common language scarcely sanctions – the more obvious form of expression being that all Zs are Xs.

I will meet the last objection first by observing that the form in which we choose that any conclusion should be presented is simply a matter of choice. Assuming the proposition The Zs which are not Xs do not exist we can at once by an application of the method reduce this to the form All Zs are Xs or to various other forms. But what is important to be noticed is that all forms which are *legitimate in themselves* whether sanctioned by the customs of ordinary language or not are in reality derivable from necessary propositions by that peculiar [C 57.22] genesis by means of the categories which have been above explained. And this leads us in answer to the previous objection to remark that the necessity for the employment of more than one proposition of common language for the expression of some of the conclusions of the method proves not the abnormal character of that method but the defective constitution of ordinary language. I do not use this term defective in any injurious sense of comparison but solely to point out that common language does not contain those elements and those combinations which are necessary to express directly the limitations to which necessary propositions become subject in virtue of *premises*. There is nothing in the *essential* constitution of language to necessitate this restriction and practically, as I have already intimated, it is not injurious, but it exists.

I will close this chapter by briefly summing up the conclusions which have been arrived at.

First the intellectual procedure so far as it depends upon the conception of Class, is determined partly by formal laws of thought partly by the relation which all class conceptions whatever bear to the conceptions of existence and non-existence.

Secondly. The principles of identity, contradiction and excluded middle have their representatives among the formal laws above referred to.

Thirdly every conclusion established by reasoning founded upon the conception of class, is in the form of a necessary proposition modified by means of the four categories described above as the universal, the indefinite, the non-existent, the impossible – the nature of this modification depending upon the premises. In other words the effect of premises is only to change the categories of some necessary proposition connecting the terms of the conclusion.

These results relate to thought as occupied about *things* but there is also a theory or doctrine of thought as exercised upon propositions – just as in the common logic we have the distinction between the logic of categoricals and that of hypotheticals. In all *formal* respects however, these theories are the same. Only it is to be observed that the science of the forms of thought as exercised about propositions, the fundamental conceptions of truth and falsehood take the place of the fundamental conceptions of existence and non-existence in the science of thought as exercised by things. From the difference of the fundamental conceptions which constitute the ground of formal law, arises a corresponding difference in the derived portions of the science e.g. as to the constitution and ...[14] [C 57.23]

Part D

Miscellaneous Matters, Letters and Fragments

Chapter XII

On Belief in Its Relation to the Understanding

—————————————[Date unknown]—————————————

As it is impossible for us to define the elementary sensations and emotions of the mind, it is equally impossible for us to define its elementary acts and operations. And as all the knowledge that we can express of the former is confined to their relations, so likewise is our knowledge of the latter. We may indeed *feel* what it is to will, to believe, to admire, but we cannot in language describe this consciousness. We can only state the general laws and relations to which those several acts and emotions of the mind are subject with reference to themselves individually and with reference to each other.

Accordingly what I propose in this essay is not to investigate the nature of that mental act which we designate belief, but to state some of its general relations. More particularly I design to consider the relations of belief to the understanding.

Every act of belief has reference to a proposition as its subject. We cannot believe without believing a proposition. That proposition is a statement, either true or false; an assertion that something exists or that some things which exist have a certain relation to each other, or lastly that some other propositions have a certain dependence upon each other. It either simply asserts the existence of an object, or it asserts a relation existing among objects, or it asserts a relation or dependence among other propositions. There is no other way in which a proposition can have meaning. It will suffice on the present occasion that we consider the second kind of proposition above described, *viz.* that which asserts the existence of a relation among objects. This is the more usual kind of proposition and all the remarks which are made with respect to it will, with slight modification, be applicable to the two other kinds of propositions referred to.

With respect to this kind of proposition, it may then easily be shewn, that whatever may be the nature of the relation which it asserts the proposition may always be reduced to a form in which it shall express the connection of two terms by the copula *is* or *are* or by some other tense of the substantive verb to be. Thus by the proposition *Caesar conquered Gaul,* we mean *Caesar was the conqueror of Gaul.* By the proposition no men are perfect, we mean *All men are not perfect,* i.e. imperfect and so on in other and more complex cases.

Now as we always know what is signified by the substantive verb which forms the copula, it follows that our under- [E3.1] standing of a proposition will depend upon our understanding of the terms of that proposition, or, to speak more definitely, that the sense in which we understand a proposition will depend upon the sense in which we understand the terms of that proposition.

For example, a child who has been instructed in that conception of God which is founded upon his attributes of power and of goodness is informed that God is a spirit. If the meaning of the term spirit has never been explained, the following will be something like the state of the child's mind with reference to the proposition in question. The child will believe that God exists and will have a distinct though necessarily inadequate conception of his attributes of goodness and power; he will believe that there exists also a something to which the name "*spirit*" is given and will believe that God is that something, or that God partakes of the qualities of that something. It may be said that in this case the proposition is to the mind of that child little if anything more than a *verbal proposition.* If more than such it is because the child believes that there *exists a something* of which the name is "spirit". Probably he will endeavour to form a material conception of that *something,* will invest it with form and personality and will make of it a picture in the imagination. But with this part of the mental process which may be supposed to be performed we are not here concerned. We confine ourselves to what may lawfully be collected from the proposition itself. As the manner in which the proposition is understood depends upon the manner in which its terms are understood, so the manner in which it is believed depends upon the manner in which it is understood. And hence, by obvious inference, the manner in which a proposition is believed depends upon the manner in which its terms are understood. In other words the state of the mind with respect to the belief of a proposition depends upon the state of the mind with reference to its understanding of the terms of that proposition.

If on the other hand the child has been instructed to form a conception whether correct or not of the meaning of "*spirit*", the proposition will no longer be to his mind a merely verbal one. He will form distinct ideas of the objects to which the terms of the proposition refer. He will attribute to those objects reality and will perceive the existence of a relation between them. To the mind thus prepared to apprehend it, the proposition may be said to be real.

As we then distinguish propositions into real and verbal according as we understand the meaning of their terms or do not, let us distinguish belief into real and verbal according to the nature of the proposition upon which it is exercised. We may then lay down the following principles.

1. That we really understand a proposition when we are able to form a clear conception of the meaning of its terms.

2. That when we cannot form a clear conception of the meaning of the terms of a proposition it expresses to us a verbal relation.

3. That we can only believe a proposition in the sense in which we understand it and therefore can only be said to have a real belief of propositions which we really understand.

Of course when we speak of understanding a term we do not [E 3.2] mean thereby a knowledge of all the properties and qualities of the object which that term represents. It is sufficient that we know some one quality or property which may serve as a mark or note whereby the object is mentally distinguished.

Granting these positions let us consider how they will affect the commonly received doctrine that We can only believe that which we understand.

It will be evident that the state of the mind cannot be such with reference to a proposition which we do not understand as with respect to one that we do understand. And it will appear from what has preceded that we cannot have a *real belief* of such a proposition in the sense in which *real belief* has above been explained. The manner and the degree in which such a proposition can be believed may be thus stated. We can believe that the terms which we do not understand express *realities of some kind*, we can believe that among those realities there exists that substantive relation which the proposition asserts by its copula *is* or *are*. I conceive therefore that it cannot with truth be said that we can have no sort of belief of a proposition which we do not understand. Our belief may amount to little more than a belief in a relation of words; but, so far as it goes, it may be a genuine act of the mind not distinguishable so far as the mental act of believing (apart from its subject) is concerned from any other act or belief.

In all true Science however, and therefore in Theology so far as it partakes of the character of a Science, it may without fear be affirmed that no merely verbal propositions, and therefore no merely verbal beliefs, can hold a place. Whatever submission of the understanding God may require of us he does not require that we should believe in the truth of propositions which we do not understand. Not even in that modified sense in which, as above said, the belief of such propositions is alone possible. It cannot however be denied that Churches have required this; that they have not seldom attributed a special merit to an unintelligent belief. Partly this has originated in a confusion between verbal belief and real belief – between the *believing that certain words express a true proposition and the believing the fact which that proposition asserts.* But it has still more originated in an unfounded notion that between Faith and Reason there is an irreconcilable opposition, and that every victory obtained over the latter redounds to the honour of the former. – It would also appear from what has been said that the distinction which has been maintained by some acute writers between unintelligible propositions and propositions concerning unintelligible subjects is unreal. In fact as a proposition merely connects terms by the copular relation *is, are* etc., it can only become unintelligible through its terms becoming so. If either of those terms be to us a mere unmeaning name, if it express a something of which we know no single quality or distinguishing mark, the proposition will be to us unintelligible, or, at the most, intelligible only in that sense which has been explained in the previous illustration. I see no other way in which a proposition can become unintelligible. It may be false or even *necessarily* false, a case which we shall proceed to consider presently, but so long as we can form any clear conception however imperfect, of what the terms mean we cannot say that we are ignorant of what is asserted and therefore cannot pronounce the proposition [E 3.3] unintelligible.

But it is not alone sufficient that the propositions of Science and of theology should be real and intelligible, it is necessary also that they should be neither self-contradictory nor contradictory of each other. They must individually be self-consistent and in their collective regard be the members of a consistent scheme. I apprehend that the consistency of any scheme of propositions consists in its being accordant with the laws of thought, or as it is sometimes said, with the laws of right reason. For in all cultivated and disciplined minds there is an agreement as to what constitutes freedom from contradiction and what does not; as to what constitutes correct inference and what does not – in short as to what is to be understood by logical consistency. It is perfectly true that we often in our unreason violate this principle, but we are nevertheless so formed that we can, by due care and attention, perceive when it is violated, and when it is regarded. And the more careful and exact our mental discipline has been, the more do we feel the shock which its violation occasions.

That it is possible for us to believe self-contradictory propositions will be made manifest by a very slight consideration. We do not in fact always exercise that degree of attention which is requisite in order to enable us to perceive that they are contradictory. The proposition, two and five are eight, is a self-contradictory proposition. It is a proposition of which the falsity may be perceived when the terms which it involves are understood. But it is also possible that those terms may be understood, and yet, through defect of attention, or of some other power, it may not be seen that the proposition is false. We need not however have recourse to the dominion of number for illustrations of the principle which is here contended for. Many propositions in which number does not enter may be seen from their implicit contradiction of some plain axiomatic truth to be necessarily false. And it must be remarked that every axiomatic truth is but a manifestation of some internal law of thought. Thus from their inconsistency with laws of thought, propositions may be seen to be false which nevertheless, that contradiction were not seen, might have been believed to be true. Now what is meant in the principle above asserted is that no proposition of this kind, no self-contradictory proposition, can hold a rightful place in the domain of Science or of Scientific Theology.

It is also asserted in the principle above referred to that no mutually contradictory propositions can together hold a place in any scheme of scientific thought. It would be unnecessary to dwell upon positions so obvious were it not that the error which is indicated is both possible and frequent. Nothing is in fact more common than for men either not to reason or to reason falsely. And in either of their cases, they are liable to the danger of believing, at the same time, propositions which are logically incompatible. It is perfectly possible notwithstanding all this, that their understanding of the import of either proposition may be complete and satisfactory. It may be the connection only which they fail to perceive.

The conditions then to which the individual propositions which are involved in any system of Scientific truth are necessarily subject are briefly the following. They must be real and intelligible— they must be self-consistent — they [E3.4] must be mutually consistent.

That it should be possible for men to believe self-contradictory and mutually contradictory propositions is a remarkable circumstance. That men should in any case be able to reason falsely is remarkable. What makes it chiefly so is that the true laws of thought, i.e. the laws of correct inference are not less rigid, not less exact, than are the laws which govern the physical universe. In one respect they indeed closely resemble the latter; they are in their ultimate form and essence mathematical. There is no example of argument or of demonstration, so complex in its form, or so remote from whatever is physical or material in its object that it cannot be conducted by a mathematical process and exhibited in a mathematical form. The common notions of mankind however associate that which is mathematical with a certain mechanical necessity. Where then shall we seek for the solution of that apparent contradiction? How is it that while the laws of correct thought resemble in their subjection to mathematical rules the laws of external nature— the former are perpetually being broken, the latter never. I apprehend that but one answer can be given to this question. It is that in the contrast here presented we behold one phase of the characteristic phenomenon of human liberty; viz. that phase which it exhibits in its relation to the intellect; just as in the moral liberty of man we survey the same phenomenon in its relation to the affections and the will. [E3.5]

Chapter XIII

The Philosophical Idea of Freedom

————————————— [Possibly 1850] —————————————

Infinite evolution

*The more complex any system is, the more difficult is it to determine its motion under external influences.

The motion of an atom simply expresses the nature of the forces to which it is subject.

It would be most of all difficult to determine the motion of a system which should contain secret springs or sources of power apt to come into operation under the influence of particular causes, yet until this influence is exercised giving us notice of their existence.

If in such a machine the springs were *infinite* it would be impossible absolutely to predict its movements at a given epoch; yet we should know that they were partly the result of external causes partly of its own internal and infinite nature. Human freedom may be something of this kind. It may be the actual development of our intellectual and moral nature, may be the unfolding of a system in which the divine wisdom has implanted infinite [B 77.1] faculties capacities emotions and seems destined to influence them but not to overbear them, not to produce the same result as if they were not.

And together with the infinite faculties capacities emotions, there may be a design in the very constitution of the system, a pervading reference to some end, in the intellect truth, in the moral part of our nature virtue in the

---

* [Headnote] Particular remarks on this freedom. Such a freedom if it be granted, analogous to but not identical with is resulting from the freedom of the will.

sensitive happiness, giving to the actual constitution and circumstance of our nature a certain teleological character.

But whether this be so or not the fact of the existence of mathematical laws in the human intellect, the fact that they are not necessarily obeyed, that their dominion in truth and not in necessity is established and if the analogy of our mental state remains unbroken the same existence of law invested with the same character of rightful but not necessary dominion must fully be admitted as our moral constitution.

If there are conclusions of science they ought to affect human conduct. Yet the prevalent feeling among intellectual men of the world seems to be that there are no characteristics of immutable laws apart from the [B 77.2] acknowledged realm of necessity.

If it were as firmly believed that truth in the intellect that justice and mercy in the life form as real a part of the design and constitution of our nature as the law of gravitation does of the material universe we should think that some influence would be produced by such a conviction.

The above in two chapters: 1st Philosophical 2ndly Historical.

Philosophical: 1st Resumé of the laws. 2nd nature and character of the laws, 3rd Relation of the mind to them.

Historical

1st cosmopœa[1] of the forms of thought: unity, duality.

Both Anselmian theory and

2nd Modern views as to the extent in which the constitution of the mind influences the forms of philosophical speculation. Question of *general ideas*. [B 77.3]

Chapter XIV

Note [to Aristotle]

The whole connected passage Metaphysics Bk 3[1] Sec. 3,4 is as follows.

3. We must moreover say whether the enquiry concerning the so-called axioms in mathematics and the enquiry concerning Being, belong to one or to different sciences. It is manifest that the speculation concerning these things belongs to one science, – and that the science of the philosopher. For these things relate to all existences and not to any genus of things separately from others and all make use of this principle that whatever exists may be considered in respect of *existence* and that each genus of things partakes of *being*. And of this principle they make use so far as is convenient for them to do – and that is so far as the genus concerning which they make their demonstrations offers occasion. Wherefore as it is manifest that the properties of *pure being* belong to all things (for this is common to them all) so the speculation of him who would recognize being as being must take account of these. Wherefore none of those who view things only in part as the geometer and the arithmetician undertakes to say anything respecting these (properties of pure being) as to whether they are true or false. Some of the physical philosophers however *perhaps* do this. For they alone supposed themselves to extend their views to all nature and to Being. Since however there is a higher than the physical philosopher (for physical nature is but a genus of Being) so there is an enquiry[2] which has reference to the Universal and to the speculative consideration concerning the Primary Existence. There is indeed a *physical* wisdom but it is not the first wisdom.[3] As to what some say concerning *truth* and the manner of receiving its evidence they speak [B164.1] falsely through ignorance of analytical principles. For they ought to approach this inquiry having a previous knowledge of those principles and not as learners. That it is the business of the philosopher and of him who would speculate concerning the origin of

all being, also to examine carefully the principles of reasoning is
manifest. It becometh him who would have some assured knowledge
of any genus of things to be able to assert what are the most certain
principles of that matter as also him who would have an assured
knowledge of existences as existences to be able to say what are the
most certain principles of all things. And this man is the philosopher.
But that is the most certain principle of all (principles) concerning
which it is impossible to be deceived. For it is necessary that such a
principle should be the most able to command assent (for all are
liable to be deceived about things which do not command assent) and
free from hypothesis. For whatsoever principle it is necessary that he
should hold who would understand existences that principle cannot
be an hypothesis. And whatsoever it is necessary that he should have
knowledge of who would have knowledge of anything it is certain
that in the possession of that principle he must approach the inquiry.
And that a principle possessing these characteristics is the most
certain of all is manifest. What it is after these things we (now) say.
*It is impossible that the same property should both belong and not
belong to the same thing at the same time.* Whatsoever other things
we would further define let us define them by referring them to
logical oppositions [i.e. to the possessing or not possessing of given
qualities].[4] But this is the most certain of all principles. For it
involves the aforesaid definition by opposition. For it is impossible
for any [B 164.2] one to suppose the same thing to be and not to be as
some think that Heraclities asserts. For it is not necessary to
suppose whatever anyone says.

For if it is not possible that *contrary* qualities should belong to the
same thing (and let us in this proposition effect the accustomed
analysis by opposition) and if an opinion is *contrary* to the opinion
which contradicts it, it is manifestly impossible that the same person
should [suppose] that to be and not to be are the same. For he who
should be deceived concerning this would hold contrary opinions.
Wherefore all who demonstrate refer their demonstration to this
ultimate opinion (axiom). For by nature it is the principle or
foundation of all other axioms.

4. Some however ignorantly think that they can demonstrate this
axiom for it is ignorance not to know of what things we may seek a
demonstration and of what things we may not. *And it is impossible
that there should be a demonstration of all things without exception,
for the chain of demonstration would extend back without limit so
that there could not thus be a demonstration.*

From this passage it appears not only that the axiomatic truth which we
have deduced as an interpretation of the fundamental law of thought

$$x(1-x) = 0$$

was recognized by Aristotle as the first and chief of axioms, but that it was also seen by him to involve the true principle of logical analysis by dichotomy, i.e. by referring to every subject the possession of one or the other member of each pair of contrary qualities. Of each subject we may thus say either that it is white or that it is not white either that it is man [B 164.3] or that it is not man and so on indefinitely. We see also that Aristotle fully recognized the necessity of some undemonstrable axioms like the above as foundation of all reasoning, a view to which his writings in almost every part bear full testimony.

The result which the researches in our text would seem to sanction is that all true axioms whatever are but the necessary interpretation of the laws of the human mind – not truths innately stamped upon the mind without reference to its constitution, but immediate results of the conditions under which its operations are performed. This is in itself a reasonable view. Nothing is more likely than that the mind should be subjected to some conditions and that these should determine the existence of some necessary relations among its conceptions. There must however be a great difference between a science constructed upon axioms and a science built upon the primary laws of thinking. The former must require the direct application of the axiom in every step of the process of reasoning the latter will embody its laws if we may [be] permitted the expression, in general methods and thus become as universal in its actual processes as in the laws upon which they are based. [B 164.4]

Chapter XV

Philosophy of Mathematics

---------------------[Probably after 1855]---------------------

The questions concerning the nature and philosophy of mathematics and which are familiar to all acquainted with the history of speculative opinion, form, it is probable, but a small part of those which have at different times occurred to earnest students without perhaps ever passing out of the silent region of thought into the outward world of controversy. The common notion of mathematics that it consists of a few clear sharp axioms, capable of being applied in infinitely various combinations and involving in the application no higher intellectual difficulty than that of arranging the order of those combinations with a view to particular ends, was never quite true and is now very widely removed from truth. Beside the metaphysical questions which form a dark background to the whole subject and which seem to have chiefly attracted the attention of those who have surveyed it from without there are others relating to the logic and philosophy of mathematical processes of which only those who study it from within can feel the difficulty and importance. Of the questions which have been most prominent whether from their intrinsic interest or from accidental association some have so changed with the progress of human thought as either to have lost all significance or to seem to us but as examples of that power of anticipation which gifted minds sometimes possess. The reality of a connection between Number and the constitution of the Universe is now among the most established truths – but we can never think of that connection in the almost material form in which it was [B123.1] thought of by the followers of Pythagoras. Other questions there are scarcely less ancient, the form and statement of which have undergone but little change. Such are those which relate to the nature of the objects about which Geometry is concerned and the source of that peculiar certainty which belongs to its primary truths – a certainty which we endeavour to describe when we term those truths necessary. These are questions of which the present interest differs but little from that with which we study the old discussions of them in Plato and Aristotle. It is a metaphysical interest which seems quite independent of the state of progress of mathematical science in the world. Again there are questions which directly depend upon that state. Every great advance in mathematics has either raised or revived such questions. For instance the discussions about ratio and proportion which occupy so large a part of Barrow's Mathematical Lectures[1] and of which the place they occupy there is only an index of that which they occupied in the thoughts and writings of others in his day may perhaps be considered as occasioned by that partial unsettling of old notions which accompanied the rise of Modern Algebra. The rise of the Differential Calculus was followed by controversies which still retain some of their old interest. Nor indeed in one sense can that interest ever pass away. He who would intelligently use that great instrument must if he do not historically follow the controversy yet work out its legitimate issues for himself. Perhaps this is the case with respect to all really great questions which have ever been subjects of controversy. Perhaps we can never fully enter into their results without in some degree going over the old ground [B 123.2] ourselves – though the position into which by the progress of time and knowledge we have been brought may by relieving us from the burden of old prejudices make easy that task which was once so hard. Modern times also have brought their own difficulties in connection with the study of mathematics the most important of which seem to be connected more or less directly with the great question of the connection of Thought and especially of Thought as occupied in Reasoning with Language. Particular applications of this question are met with in the various discussions which have arisen respecting the use of "imaginary quantities", the validity of the methods called symbolical and kindred subjects.

It must at first sight appear remarkable that in the realm of mathematics controversies some of which have been perpetuated from age to age should continue to exist. And perhaps it does thus appear strange because of the contrast which they present with the great external facts of the history of the Science, its career of majestic conquest, the consent invariable and universal which its conclusions have claimed and have received. Nor can the question fail to suggest itself: Must this contrast exist always? Is it in the nature of things that while the history of Mathematics is in one sense that of an unbroken career of discovery it is in another sense a history of perpetual doubt and conflict? Now it does not seem clear that any general answer can be given to this question. The points about which controversy has arisen are as it has been seen of such different kinds that a probable opinion of its issue in one class of cases would throw little or no light upon the others. We must consider the [B 123.3] different classes separately. This is indeed a considerable part of the design of the present

work – but there are some considerations on the different parts of the subject
which may with propriety be stated here.

1st. A careful study of some of the more prominent of the views which
have been held of the philosophy of mathematics will probably favour the
conclusion that some of those which appear to be in more direct antagonism
are not so much erroneous as partial – that they need only to be completed
each by the truths in which it is defective in order to come into mutual
agreement. All theories are attempts to realize a speculative unity and
among the possibilities of failure this must be reckoned, viz., that they may
aim at a unity which is too narrow. It has appeared to me for instance that
this is the case in some of the more important theories which have prevailed
as to the nature of Geometry. Each of them contains some partial truth
which through a too eager desire for unity has been magnified into undue
proportions, and made to stand for the whole. One writer would build
Geometry entirely upon definitions, another upon axioms, another upon
Postulates. Again with relation to the Logic of Geometry, one writer regards
it as an application of syllogism a series of steps of deduction each of which
consists in inferring the particular from the general, a second maintains that
all its steps are from the particular to the particular, a third arriving at a
more absolute unity can see in the entire science nothing more than an
application of the dictum that whatever is is (principle of identity). These
latter differences of opinion have it is true not *originated* in speculations
about Geometry but they quite as much belong to the general question of the
Nature of Geometry as do the more peculiar inquiries respecting [B 123.4]
the place and office of the Definition the Axiom and the Postulate. In all
these views there seems to me to be a measure of truth – though in some of
them the truth which is retained is but small in comparison with that which
is excluded. But whatever scattered truths they may contain no theory can
claim acceptance which does not assign to those truths their several places
and proportions. The unity after which we are bound to aspire is not that
which consists in putting a part for the whole but that which determines and
prescribes all the relations of parts.

2ndly. Some of the questions which have been most controverted are as
has already been stated of a metaphysical nature – and the solutions which
they receive in the individual mind will usually differ according to its
particular leaning or bias in questions of philosophy. Thus according to the
degree of relative importance which is assigned to the sensational or to the
ideal elements of knowledge it is probable that the judgment will be
determined with reference to some of the questions above noticed e.g. the
nature of the objects about which Geometry is concerned and the ground of
that peculiar certainty which attaches to its conclusions. We might predicate
beforehand that upon such subjects the views of Plato would have but little
affinity with those of Hobbes. Now it may be asked what if such be the case
is the use of discussion [of] such particular questions at all seeing that the

answers to them are necessarily implicated in more general views. To this I would reply that if on the one hand our judgment on particular questions in the philosophy of mathematics is likely to be influenced by the general tendencies of the school of philosophy to which we may belong, on the other hand there is no particular class of questions [B123.5] which if independently studied is so likely to modify or even to determine our general views as this. It is possible that the whole of a man's general views of philosophy or at least of that part of philosophy which relates to the theory of knowledge may be but the consistent development of principles which he has derived from reflection upon the grounds and the processes of mathematics. Certainly there is no other class of subjects to which the great masters of philosophy have so often appealed – none from which they have derived illustration so apt and clear and exact.

These metaphysical difficulties have as might be expected no effect upon the progress of mathematics as a science. Whatever view we may take of the nature and ground of the certainty of the primary truths of Geometry that certainty admits of no question. The object of disputation indeed is not to establish but to explain a fact. It does not cease on this account to be important that this fact should be explained and whether it be important or not we are so constituted that we can scarcely help endeavouring to construct some kind of theory for its explanation.

3rdly. There are however differences of opinion; and where absolute differences of opinion do not arise there are often uncertainty, looseness of comprehension and want of clear definite thought among even the most advanced students of mathematics upon questions which cannot strictly be termed metaphysical and which do in a very important degree affect the progress of the science. This seems to be especially the case with respect to that entire class of questions already referred to which clusters about the great subject of the [B123.6] dependence of reasoning upon language. Perplexities of this kind must have been of early origin in all those parts of mathematics, including common Algebra, which are known under the name of Analysis, and they still form a very considerable part of the difficulty which belongs to the higher parts of the Science. Questions respecting the kind of interpretation which must be given to a formula, and limits within which such interpretation is valid – the conditions under which methods and processes may be lawfully applied, are of very frequent occurrence – And it is not likely that so long as mathematics is a progressive science such questions will ever cease to present themselves. Till those limits are reached which either the nature of mathematical truth or the nature of the human faculties imposes an end of controversy is not to be looked for.

At the same time there seems to be no reason to think that the number of really great principles which is or can be involved as matter of controversy affecting the actual progress of mathematics can ever be considerable. The probability seems to be that it is and always must be very small. Nor again does experience shew that controversy can settle for any great length of time upon any question of the absolute truth or falsehood of principles. Perhaps all really important principles are so far axiomatic in their character that there exist particular cases in which they are seen with something like the self-evidence of axioms. The reason why it is possible for them to be again and again the subjects of debate if not in a public manner yet in the individual mind, is that no particular cases can exhaust their meaning, no finite form [B 123.7] of words can give to them adequate expression. We do not indeed find it difficult to give to them an expression adequate for the particular stages of progress and attainment, but new cases arise for which the ground of enquiry must be gone over afresh. It may be that when this has been done we find that the principle involved is essentially the same as that with which we were acquainted before. But our acquaintance with it is no longer the same. If we still retain the old formula for its expression we attach to the terms of that formula a wider and a deeper meaning. The kind of change which takes place in the mind of one who is continually advancing in real mathematical knowledge is not so much that he is continually acquiring new principles, as that he is continually growing up if I may use the expression into fuller comprehension of the meaning of the old. The larger questions which meet us seem to admit only of what may be termed a *progressive* solution.

The considerations which have been stated will indicate with sufficient clearness the general design of the present treatise. It is proposed to examine in succession some of the most important classes of questions relating to the Philosophy of Mathematics beginning with Geometry and ending with those methods and processes of the higher Analysis which are the least dependent upon any visible representations.

Chapter II: On Geometry

In the present chapter it is proposed to describe with brief comment some of the more obvious characteristics of Geometry as a science. We choose as the more immediate [B 123.8] subject of examination the first six books of the elements of Euclid. In other chapters we shall enter more fully into some of the questions which will be suggested in this.

As presented by Euclid Plane Geometry is a system of reasoning, and of truths deduced by reasoning, about figures described upon a plane surface the elementary parts of each construction being straight lines or circles. The figures are presented visibly as the direct objects of reasoning; their elementary parts as well as the more important construction formed by their combination are *defined*, the conditions under which the constructions are made are stated in Postulates, and the elementary general truths of which the reasoning is an application are expressed in *Axioms*. In these particulars, – the visible representation, the clear definition, the constant reference to postulates and to axioms, the peculiar value of Geometry as a discipline has always been felt to consist.

We shall first speak of the definitions.

The main purpose of a definition is not description but distinction. The thing to be defined must be distinguished from all other things – and for this end the distinguishing property is described. It is possible that there may exist more than one property or set of properties that would suffice for distinction. In this case that one is chosen which is thought to be either most simple or most fundamental, or which for particular reasons is thought most deserving of attention. The name triangle implies that the figure denoted has three angles – and this with the implied condition that the sides are rectilineal would form a definition. But it is simpler and at the same time more fundamental to define the figure as [B 123.9] bounded by three straight lines – because the notion of a straight line is prior to that of a rectilineal angle. Again there is a quadrilateral figure which might be defined either as having its opposite sides parallel or as having its opposite sides equal, Euclid takes the former definition simply and calls the figure parallelogram but he also takes the latter and distinguishing the cases in which the sides are equal from that in which they are unequal calls the figure in the one case a Rhombus in the other a Rhomboid. In these added definitions the object is not so much distinction as the directing of the attention to a particular property rather than to another as a ground of deduction.

We see in this example that definitions are in some measure arbitrary. And as there is perhaps no question of more fundamental importance than to determine to what extent they are arbitrary I purpose to say a few words on this point here. The whole question will again and again recur.

Definitions are certainly arbitrary to this extent viz. that among *possible* constructions we may select those which we please as the objects of definition. A triangle is a possible construction and it has been rightly chosen as an object of definition. Among triangles the equilateral the isosceles and the scalene have been selected as objects of more particular definition – so have the species of triangles called right angled, acute angled and oblique angled. In the one case attention is directed to distinctions dependent on the relations of sides; in the other to distinctions dependent upon the magnitude of angles. This process of particular definition might be carried on a great deal further – and each step [B123.10] would be arbitrary so far as it involved an act of choice directing us to particular points of construction and not to others but not beyond this. The far more important question What constructions are possible involves nothing arbitrary. In every definition there is something which is not founded upon hypothesis.

For instance, the definitions of the different rectilineal figures rest all upon the common grounds of the postulate that it is possible to draw a straight line from any one point of a plane to any other point of the same plane and upon the axiom that two straight lines cannot, but that more than two straight lines can, inclose a space. Euclid does not express the positive portion of this axiom but it is implied in his statement of the negative one. The definition of parallel lines involves without doubt the unexpressed postulate that it is possible for two straight lines to exist which if continued indefinitely would never meet. Our conceptions of geometrical figures are not arbitrary – and though we speak and truly, of the infinite variety of constructions, yet is that variety but the combination of elements which are in themselves determinate in nature and relation.

What has now been said will be sufficient to shew us the erroneous character of that view which represents geometry as entirely founded on definitions – and its theorems as merely the consequences of particular hypotheses. It is forgotten that though in the expression of geometrical propositions we often employ the language of hypothesis e.g. If a figure be constructed under such condition it will possess such and such properties; yet the conditions themselves have a real foundation in the [B 123.11] nature of things i.e. either in the absolute nature of space itself or in the relations by which space and its affections become objects of human thought. When the demonstrative character of Geometry is said to rest solely upon the fact that it deals with hypotheses we are in danger of supposing that it is a matter of human and of arbitrary creation and of putting out of sight the very fact which it is sought to explain.

We are thus led to the threshold of the important question of the *reality* of the objects about which Geometry is concerned. It is no uncommon view which represents Geometry not only as founded upon hypotheses but upon hypotheses which are nowhere in the world actually realised. The perfect straight line or circle it is said has no existence and cannot even be conceived.

To this it has been well replied that assuming the external reality or to use a term familiar to metaphysicians the objectivity of space, geometrical figures exist in space even as they are defined. To adopt the striking language of Mr. Hallam, every geometrical figure "exists in the infinite round about us as the statue exists in the block". "No one can doubt," he continues, "if he turns his mind to the subject, that every point of space is equidistant, in all directions, from certain other points. Draw a line through all these and you have the circumference of a circle (sphere); but the circle itself and its circumference exist before the latter is delineated. Thus the orbit of a planet is not a regular geometrical figure, because certain forces disturb it. But the disturbance means only a deviation from a line which exists really in space, and which the planet would actually describe, if there

were nothing in the universe but itself and the centre of attraction."[2]
[B123.12]

In these observations I fully concur. Granting the existence of space the
figures of geometry *exist*. The degree of perfection in which we can realise
them does not affect their existence nor, if we consider the matter carefully,
does it affect our conviction of their existence. If in comparing two attempted
constructions of the same figure e.g. the circle we can upon the testimony of
the senses affirm that the one is more perfect than the other it can only be
because the ideally perfect figure exists not merely as a conception in our
minds but as a reality in space – a reality approachable though not
attainable. The language ordinarily employed on this subject would seem to
recognise the existence of imperfect figures in space but not of perfect
figures. Now it is well to observe that the difficulty arising from the
imperfection of the senses affects the one just as much as the other. That
difficulty does not consist in a simple inability to construct the perfect figures
of geometry, but in an inability either perfectly to reproduce or perfectly to
imagine any figure whatever. I here use the word imagine to express that
power of the mind by which we *picture* to ourselves a thing or try to think of
it as it would *appear*. Now this power is in its very nature imperfect and
limited – imperfect because it aims at most but to reproduce the impressions
of our imperfect senses – finite and limited because subject to conditions to
which that higher power of the understanding by which we apprehend
relations is not subject. For instance judging from personal experience I
should say that we are not able to *imagine* length absolutely without
breadth, and so to *picture* to ourselves the perfect line of Euclid. We
approach it when [B123.13] we think of the edge or boundary of a surface –
but here it is still associated with the idea of breadth. But though we cannot
imagine pure length, we can both conceive its existence in the infinite round
about us and can understand its relations. Now it is this understanding of its
relations or to speak more precisely of the fundamental relations involved in
the definition of the straight line which constitutes the scientific conception.
The material form, the mental image, are important only as they suggest
those relations.

These considerations lead us to apprehend the twofold truth that while
the perfect figures of Geometry have a real existence in space they are to
imaginations striving to realise them as they are but the *limits* of an
indefinite process of abstraction. There is no contradiction here. Our power
to imagine or to construct is not only no measure of what is possible in the
nature of things but what is more remarkable it is no measure of our power
to apprehend relations. And I think it deserving of notice that the inability of
the senses and of the power of imagination to represent to us the objects of
Geometry as they really are carries with it in some degree its own correction.
That imperfection of faculty which prevents us from reaching the limit
renders us at the same time insensible to the degree of our shortcomings.

The constructions which satisfy us now might, were the faculties by which we estimate measure and number and direction developed to a higher degree of precision, cease to convey to us any distinct ideas.

There are certain definitions in the text of Euclid which must be considered as unsatisfactory, either from their indefinite character or from their not containing all that is really [B 123.14] involved in the conception to be defined and needing directly or indirectly a supplementary addition. Of this kind is the definition A straight line is that which lies evenly between its extreme points.* The idea conveyed is that of similarity and uniformity not only of the different parts of the line to each other but of the uniformity of each part of the line with respect to the different sides on which it can be contemplated. To make the definition formally complete something would be required equivalent to the condition that the line may be regarded as an *axis* – that two of its points being fixed it may be supposed to revolve without changing its actual position in space. No doubt any such definition as this really involves the Geometry not of the plane merely – but of space in its three dimensions – but unless the plane be defined before the straight line, which Euclid does not do, any definition of the latter must involve the consideration of space generally – and the seeming escape from this in Euclid's definition is only due to its vagueness. Actually his definition is completed so as to become susceptible of scientific application in the axiom that two straight lines cannot enclose a space.

I am indeed disposed to think that the *Natural History* of Thought in the subject of Geometry is something of this kind. We begin with that crude conception of Space which is perhaps best expressed in English by the word "room" – we advance by the teaching of the senses and the action of the abstracting mind upon the teaching to those popular and as yet unscientific conceptions which are implied by the words as surface, line, point, figure etc.; and their more obvious relations – [B 123.15] we subsequently are taught to give to those conceptions scientific form and to employ them in deductive reasoning. Now what is the mode of this transition? It is obvious that it partly depends upon that power of abstraction by which we can separate the different elements of thought which are involved in the more immediate images of the senses and intellectually contemplate them apart. I do not mean by this that we have the power to perfectly picture to ourselves by any effort of thought a line according to its formal definition as length absolutely without breadth – but we have the power to approach that conception even as a mental image, to contemplate it as a *limit* to which the power [of] imagination can indefinitely approximate – and still more to affirm those propositions which in the limit are absolutely true and which in the world of matter approach the more nearly to realisation as the limit itself is the more

---

* Εὐθεῖα γραμμή ἐστιν, ἥτις ἐξ ἴσου τοῖς ἐφ' ἑαυτῆς σημείοις κεῖται.[3]

nearly approached. On the nature of this conception and on the reality of its object we shall speak more fully hereafter. As another example, we learn from experience to distinguish between *form* and *position*. We see the same object change its position without altering its form. It may be true that we cannot picture or image to ourselves form without position. But that the elements are distinct, that relations dependent upon form remain unaltered by mere change of position, is a truth which is felt and acted upon before it is contemplated as truth. Now this truth takes its true scientific place in Geometry in certain principles admitting of precise definition and application e.g. the principle of superposition affirming that figures which by a supposed change of place can be made to coincide are equal. [B 123.16] It is true that this principle is not in so many words affirmed by Euclid; but it is the real ground upon which he makes use of his eighth axiom viz.: Magnitudes which coincide with one another, that is which exactly fill the same space are equal to one another.* Gradually therefore the popular and somewhat ...[4] conceptions which we derive from experience are purified from foreign admixture and are seen in their true measures of independence or relation.

Perhaps the most natural development of the conceptions of the plane and the straight line would be the following viz.: [1st.] To define the plane as the surface which lies ἐξ ἴσου[5] between the two portions of space which it separates i.e. so that the one portion might be made by change of position to coincide with the other, the plane *reversed* occupying the same position as before. 2nd. To define the straight line as the line on a plane which lies ἐξ ἴσου between the two portions of the plane which it separates. I think this mode of consideration would express most directly the ordinary and popular conceptions of the plane and straight line – of which symmetry on opposite sides certainly forms a part.

But whatever modes of representation we adopt the general principles on which the definitions of Geometry rest remain the same. They are expressions in a form adapted to scientific use of the fundamental conceptions of space and of its affections which we arrive at by the abstracting power of the mind operating on the experience afforded by the senses, and which though involved in ever changing products of experience are themselves not arbitrary or subject to change.

We have next to speak of the Axioms and the Postulates. [B 123.17] In the ordinary translations of the Elements of Euclid the postulates are only three in number, viz

---

* Καὶ τὰ ἐφαρμόζοντα ἐπ' ἄλληλα ἴσα ἀλλήλοις ἐστίν.[6]

1. Let it be granted that a straight line may be drawn from any one point of space to any other point.

2. And that a terminated straight line may be produced to any length in a straight line.

3. And that a circle may be described from any centre and at any distance from (with any radius about) that centre.

But in the Greek text of Peyrard,[7] under the same head of Postulates (Αἰτήματα)[8] occur also the three following

4. And that all right angles are equal to one another.

5. And that if a straight line meet two other straight lines so as to make the two interior angles on the same side of it taken together less than two right angles these straight lines being continually produced shall meet on that side on which are the angles which are less than two right angles.

6. And that two straight lines cannot enclose a space.

The last three, it is well known, occur in the ordinary translations among the axioms. But a little attention will shew that the place which they there hold is peculiar. They are in a special manner *geometrical* axioms. The others with the single exception of the axiom "Magnitudes which coincide with one another are equal to one another" are not in any peculiar sense geometrical. The truths that the whole is greater than its part that the doubles of equals are equal etc. apply to the conceptions of geometry only because and in so far as these conceptions fall under the more general conception of *quantity*. They have nothing to do with *figure or position*. The very term "common notions" (κοιναὶ ἔννοιαι)[9] by which they [B123.18] are designated indicates that they are not confined to Geometry alone. And they are ordinarily recognized as the axioms which are common to Algebra and to Geometry.

We see then that while the view generally held at the present time of the distinction between Postulates and Axioms is that the former relates to the possibility of a construction the latter to the necessity of a truth, the former presenting the same analogue to a Problem which the latter does to a Theorem, Euclid's view of the distinction was of a totally different kind. The axioms were according to him the fundamental truths or notions which flow from the very conception of magnitude – truths which are equally valid whether applied to that discrete form of magnitude which we term Number, or to the continuous magnitude with which we are concerned in Geometry. The Postulates have reference to those more special though still fundamental truths which are involved in the conceptions of space and figure. It makes no difference according to his view whether the latter are presented in the form of demanding our assent to a truth or demanding our admission of the possibility of a construction. The admission of the possibility of a construction is as much a development of our conceptions of Space and Figure as is the admission of a necessary truth about Space or Figure. It would be easy to convert the one class of Postulates into the other. Every geometrical truth implies a construction in which it is set forth and manifested; every geometrical construction is founded upon a truth.

This distinction so fully recognized by Euclid and in later times so much lost sight of is one of great importance. If we would understand the true nature of geometry we must [B 123.19] distinguish between the different kinds of principles which form its basis – we must endeavour to assign to each class its proper functions.

In the formal development of the Science of Geometry the Axioms usually constitute the major or universal propositions of Syllogisms – the minor or particular propositions of which express the particular conditions of the given construction. And those postulates which are not presented in the forms of propositions but which demand our assent to the possibility of certain general constructions may be considered as the suppressed major premises of syllogisms, the conclusions of which affirm the possibility of the actual construction of the problem or demonstration in question. In this way it becomes possible to present the entire procedure of the Science in the forms of consecutive Syllogisms. And the system of Geometry in actual use are so far constructed upon this type that their successive steps are syllogisms either in form or intention – the chief departure from the scholastic forms being that the enthymeme and the so-called rhetorical syllogism in which the minor premiss precedes the major, take the place of the regular Aristotelian figures. It has perhaps from this circumstance come to be considered that the reasonings of Geometry are in their own nature syllogistic – that Geometry is no more than an application to the particular matter about which it is conversant of the Logic of the Schools.

Now it is not my intention to enter in the present chapter upon the consideration of this question. The discussions of the foregoing sections have however to a certain extent prepared the way for its consideration. They have shewn us [B 123.20] by actual analysis that there can at most be but three distinct classes of principles involved in the scientific foundation of Geometry as a Science viz.: 1st. The principles of common Logic (I use this term not with special reference to the scholastic forms of the Science but with reference to its object or matter – the relations of *genus* and *species*, of general and particular); 2ndly. Principles depending upon the conception of magnitude. 3rdly. Principles depending upon the conception of Space and the subordinate conception of Figure. This classification is an exhaustive one. Whether or not we can reduce these to mutual dependence, or can establish the superior importance of one class over the others, we at least do not need to enquire for more. [B123.21]

Chapter XVI

[Various fragments, apparently late]

――――――――――――――――[Later than 1855]―――――――――――――――――

[Fragment 1]

There is perhaps no property of the human mind which is so wonderful as its power of introspection. That it should be capable of sustaining the dual relation of subject and object, that it should be able to investigate its own constitution to engage its faculties in the study of themselves is a fact which if we apprehend it at all we cannot apprehend without admiration. I speak here in a more especial manner of the mind in its intellectual capacity in its faculties of thought and reasoning as distinguished from those which are concerned with emotion and will. And within the compass thus marked out how many questions of deep interest present themselves. Is it for example possible to establish anything like unanimity of sentiment in a field of enquiry which while it lies open to every man's personal observation has above all others been fruitful in controversy? Is it possible to arrive at a just estimate of truths which Aristotle regarded as the foundation of philosophy and which Locke pronounced to be "trifling", which the pro- [A 11(b)/1] found mind of Leibnitz would within the province of Logic at least have raised to their ancient repute and which the clearest and most popular of modern expositors of the Science seems equally disposed to ignore? And if these things are possible what finally is the nature of that system of laws the knowledge of which must appear as at once a revelation of consciousness and a reward of careful and philosophical enquiry? These are questions which I propose to endeavour to answer in the present essay. The grounds of much of what I shall have to say are already before the world in a special treatise on the laws of Thought. But other results which I shall have occasion to present are founded upon more recent investigations and it is in the connexion of the whole that the grounds of the argument which I seek to develop chiefly rest. [B 7/2]

[Fragment 2]

Digression on the Nature of Algebra

In the system of ordinary Algebra we employ letters or figures to represent numbers and by connecting symbols or by relative position we express the operations and relations to which numbers are subject. Thus if x and y represent any two numbers whatever $x+y$ will express their sum, xy their product, $x-y$ the remainder obtained by subtracting the number y from the number x and so on. In like manner $x = y$ will express that those numbers stand to each other in a relation of equality. It is to be noted that while the numerical symbols of Arithmetic as 1, 2, etc. are employed only for the expression of particular and known numbers, literal symbols as x, y, though they may be used in the above sense are usually employed in a more comprehensive one viz. either for the expression of any number whatever as in the general theorems of Algebra, or for the expression of particular but unknown numbers whose values it is sought to determine. And in either of these latter uses they must be regarded as general symbols and considered as subject to those laws which are derived from the very conception of number as such. Thus whatever numbers are represented by x and y it is universally true that

$$x + y = y + x \qquad \text{(i)}^1$$

$$xy = yx \qquad \text{(ii)}$$

$$z(x+y) = zx + zy \qquad \text{(iii)}$$

to which some other laws might be added connected with the use of the symbol − (minus). If we proceed to equations, i.e. propo- [A 67/1] sitions affirming the equality of numbers whether expressed by simple symbols or by combination of symbols other necessary laws present themselves. Thus if

$$x = y$$

and
$$w = z$$

$$x + w = y + z \qquad \text{(iv)}$$

$$xy = wz \qquad \text{(v)}$$

and so on. These laws constitute the foundations of Algebra. They find direct expression in its axioms and first principles. Algebra as a developed science consists in the expression of numerical relations by general or particular symbols of number and in the application of such laws to the solution of problems relating to number or quantity.

From this conception of an algebra whose general symbols are supposed to represent any numbers whatever and whose laws are therefore founded upon the laws of number as such, we readily pass to the conception of an Algebra whose symbols are of a more restricted character. We can for example conceive of an Algebra of whole numbers exclusively, i.e. of an Algebra whose general symbols x, y, etc. should be defined to represent whole numbers generally and whose particular symbols should only represent particular whole numbers. Consider then for a moment what would be the laws of such an Algebra. They would constitute a system including and comprehending the laws of the more general Algebra before considered (because whole num- [A11/2] bers must at least be subject to those laws which are common to all numbers whatever) but also including those special laws whatever they may be to which whole numbers alone are subject. On the same principle we might conceive of an Algebra more restricted still, of one for instance whose particular symbols of number should be two only, viz. 0 and 1 and whose general symbols should admit of either of these determinations but of no others. The laws of this species of algebra would by what has been said include the laws of general Algebra together with those which are peculiar to the numbers 0 and 1 but which at the same time are common to those numbers.

Now that law which is peculiar to but common to the number 0 and 1 is expressed by the equation $x^n = x$. For if x stand either for 0 and 1 this equation is satisfied but if x have any other value it is not satisfied. We may therefore say that an algebra the interpretation of whose symbols of quantity should be confined to the numbers 0 and 1 must obey the laws of ordinary Algebra together with that additional law which is expressed by the equation $x^n = x$. Thus if x, y, z etc. be its general symbols [B8/3] we must have in virtue of the former condition

$$x + y = y + x$$

$$xy = yx$$

$$z(x+y) = zx + zy$$

vide (i)(ii)(iii) as also the laws expressed by (iv)(v), while in virtue of the latter condition it must satisfy the relations

$$x^2 = x \qquad y^2 = y \qquad z^2 = z$$

Now these laws are so far as we have determined them identical in expression with those of the symbols of Logic, vide sec. ...[2] And a more complete examination of the two systems shews that the identity is continued throughout. I desire to avoid all occasion of mistake in the statement of this truth. Let it then be distinctly observed that the literal symbols are different in meaning in the two cases; in the one they represent things, in the other a particular species of numbers. Let it equally be noticed that the operations to which the symbols are subject are not necessarily and throughout the same. The operations denoted by the symbols + − are indeed the same or closely analogous, those denoted by relative position as in the expression xy are apparently wholly different in their nature. In the one case xy denotes the product of the numbers denoted by x and y, in the other the class whose members comprise the individuals common to the classes x and y. Yet is it true in both [B6/1] cases that $xy = yx$ and that $xx = x$. The two systems of interpretation and expression are established on separate conventions; and their laws are independently determined. And this having been done it is a conclusion founded on actual comparison and which apparently could not have been predicted, that the laws of the two systems are identical in expression; we think of them or rather we think in them according to the same mental formulæ. This identity is the basis of the Calculus of Logic.

To many minds and these not undisciplined by philosophy, a difficulty will probably arise in accepting the idea which I have endeavoured to present. It will be granted that there is a sense in which the dominion of numbers is universal − at least as respects the material Universe. That all things were made in measure and weight and number is to this extent a truth not less of modern science than of ancient philosophy. But the merely numerical element − the fact that classes consist of individuals and that individuals are capable of being numbered is one with which we are quite unconcerned in Logic. The validity of the syllogism

All men are mortal   [A11o/2]

Kings are Men

therefore Kings are mortal

is wholly independent of the question whether the members of the human family and the race of Kings be many or few. All this is undoubtedly true. It is not because classes consist of individuals capable of numeration that the laws of thought are such as we find them to be. And therefore it is not in the usual and ordinary sense that we can establish any connexion between the science of Logic and that of number. But if we remember that we reason by the aid of language and that we never can express by the signs of language either the nature of things or the nature of number it must follow that it is only with the forms of thought as involved in the use of those signs that our concern lies. And there is no reason *a priori* why the forms of thought in algebra or in a particular species of algebra should not be the same as those which govern the deductive process in relation to things. On the contrary the question whether they do so or not is one possessing much speculative interest apart from any use that may be made of its solution. Now actual comparison shews that such a relation does exist. And though [A11(k)/3] a knowledge of this relation is not necessary to enable us to reason in a purely formal manner by the aid of symbols, it is apparently necessary in order to [achieve] the establishment of a *General Method* in Logic. [A11(g)/4]

[Fragment 3]

Principles of Expression

Let letters be employed to denote things or classes of, whether expressed in ordinary language by proper names as "Peter" or by nouns substantive as "men", or by the use of adjectives as "white things" or by descriptions or definitions marking out the things signified from all others.

Let the collection of two letters as xy denote that class of things which is formed by the individuals that are common to the classes denoted by x and y. Thus if x represents sheep and y white things let xy denote white sheep.

Let the symbol + written between the expressions of two classes as $x+y$ denote the aggregate which is formed by adding the individuals of the classes x and y together so as to form a whole. Thus if x represent men and y women $x+y$ will represent men and women.*

Let the sign − be used as the equivalent of the conjunction *except*, so that $x-y$ expresses the remainder which is left when from the class x we take away the class [B11/5] denoted by y. The condition of interpretability of the expression $x-y$ evidently is that there should exist no individuals in the class y which are not found in the class x.

As we are at liberty to give to symbols what fixed interpretation we please, we may employ the symbol 1 to represent the Universe or rather all things contained in the Universe and the symbol 0 to represent Nothing. I shall afterwards shew that there is a real ground for that interpretation. It will follow that if x represent a particular class as "men" $1-x$ will represent the class "not men" for reading the symbols in order we have "All things except men", i.e. the class "not-men". In accordance with these conventions

---

* It is to be observed that the expression $x+y$ is uninterpretable unless the classes x and y are entirely distinct so as to be capable of forming parts of a whole. This may be termed the condition of its interpretability. Such conditions of interpretability are really implied in ordinary language. The expression "Trees and Plants" is strictly speaking devoid of meaning unless it be understood that trees are not plants.

the expression $y(1-x)$ would represent those things which are in the class y but not in the class x. The expression $(1-x)(1-y)$ [represents] those things which are contained neither in the class x nor in the class y. Brackets are used in combination as a single symbol. A vinculum or bar over the expression may be used in the same sense.

Thus $x(1-y)+y(1-x)$ and $x\overline{1-y}+y\overline{1-x}$ would be identical in meaning. That meaning would be "Things which are in the class x but not in the class y together with things which are in the class y but not in the class x", or "Things which are either but not both". The expression $z\{x\overline{1-y}+y\overline{1-x}\}$ [A11n/2] would on the above principles be interpreted as follows: "Things which are common to the class z and to the class whose members ar either x's or y's but not both."

Laws of the Symbols

It is shown (Laws of Thought Cap. II)[3] that the symbols above described satisfy the following laws viz.

$$xy = yx \tag{1}$$

$$xx = x \qquad \text{or} \qquad x^n = x \tag{2}$$

$$x+y = z+x \tag{3}$$

$$z(x+y) = zx+zy \tag{4}$$

and to these may be added some other laws connected with the use of the sign $-$. For the nature and origin of those laws the reader is directed to the treatise to which reference has been made. As to the sense in which they are affirmed to be true let this statement suffice. Whenever such definite class-meanings are assigned to the symbols x, y et cetera as to make the two members of any of the above equations (1) (2) [et cetera] interpretable, i.e. expressive of conceptions possible to the mind, then will the individuals grouped under one of those conceptions be identical with those grouped under the other. I will [interpret] as an illustration the law expressed by (4). Let then x represent trees, y flowers and z withered things in general. Then the interpretation of the first member of (4) is "withered trees [B 165/?] and flowers", the second member as "withered trees and withered flowers". And the equation (4) connecting those members affirms that these two conceptions are identical in their comprehension, i.e. that the individual things comprehended under the description "withered trees and flowers" are the same as the individual things comprehended under the description "withered trees and withered flowers". The mental order in which the two conceptions are formed is indeed different. In the one case we connect into a single whole trees and flowers and from this whole select all the individuals to which the term withered is applicable. In the other case we form first the

conception of trees as a whole and mentally select those of the class which
are withered thus forming the conception of "withered trees", in like manner
we form the conception of "withered flowers". Now the law expressed by (4)
affirms that these different modes of forming an ultimate conception are
equivalent in the sense above explained. And all the laws which have been
noticed are of the same kind. They all affirm the possibility and the lawful-
ness of conceiving in different ways and according to a different mental order
things substantially the same. They are properly speaking laws of apprehension. [B9/4]

[Fragment 4]

And here arises a grave objection against the Aristotelian and all similar
expositions of the science of Logic. It is that while they recognise the faculty
or operation which they term Apprehension and by a special term
acknowledge in certain cases its complex character they do not attempt to
give any account of its laws. The complex does not however arise from the
simple without combination and combination is not unregulated and without
law. How the laws of Apprehension affect the form and method of the science
of Logic it is not possible agreeably to what has already been said to
determine beforehand. But it is a fact of which abundant verification will be
afforded that this influence is of a very important kind. Perhaps my meaning
will be made more clear by an illustration which I borrow from a recently
published tract of Leibnitz.* Representing since signs are arbitrary in
[B13/5] their mere form, simple names and words descriptive of quality by
letters, as

    a pro animal
    b   –   rationale
    c   –   mortale
    d   –   visibile[5]

he goes on to describe the expressing of terms and propositions and the laws
of the signs employed nearly as follows. 1) "Terminus est a. b. ab. bcd. ut:
animal, rationale, animal rationale, rationale mortale visibile.
Propositionem universalem affirmativam sic designo: a est b ubi a subjectum
b praedicatum. Est: copula."[6] And then under the head Principia Calculi he
lays down among other principles the two following viz.: "2) Transpositio
literarum in eodem termino nihil mutat ut *ab* coincidet cum *ba* seu *animal
rationale* et *rationale animal*. 3) Repetitio ejusdem literae in eodem termino

---

* "Addenda ad Specimen Calculi Universalis".[4]

est inutilis, ut *b* est *aa,* vel *bb* est *a*; homo est animal animal, vel homo homo est animal. Sufficit enim dici *b* est *a*[7] seu homo est [B 12/6] animal."[8] [In this passage *b* stands for *homo* and *a* for *animal.*][9]

A familiar example drawn from ordinary language of the law to which Leibnitz refers in 2 is seen in the *indifference* of the order in which adjectives absolute in their significance and applied to the same subject follow each other. The expression "white horned sheep" points out the same class of individuals as the expression "horned white sheep". This is immediately the consequence of a law of *language* but remotely and fundamentally of a law of *thought*. (Laws of Thought ...)[10] The same remark applies to the second law noticed by Leibnitz. If we enquire carefully into their origin it will be found that they have their ultimate ground in that faculty which Logicians term Apprehension, but which modern writers on the philosophy of the mind have usually designated as Conception. Now if these laws are contemplated merely as isolated truths they will probably appear to be of but slight value and importance. To a superficial gaze they will *certainly* appear so. But if they form, as they do, a part of a perfectly definite system of laws relating not to the faculty of Apprehension alone [A 11(d)/7] but to thought in its several faculties of Apprehension, Judgment and Reasoning, it would be an unwarrantable presumption to set them aside as if [of] no possible use or interest. And taking the larger view of the question a strong *prima facie* objection is seen to arise against the assumption of completeness in any system of Logic however strong by authority or venerable by time, in which they are actually neglected.

I shall now proceed to give some account of that scientific development of logic to which the previous remarks have been introductory.

Now every scientific development of Logic presupposes the possibility of separating by abstraction the material and the formal elements of language or to use a less technical phrase of distinguishing between the office of words and their special meaning. We are able for example to contemplate the noun substantive as such and to study the laws to which from its very office of representing *things* it is subject – laws which are independent of the nature of the things represented. As signs moreover are in their construction arbitrary we can replace words by letters, i.e. we can represent the things about which we reason by letters. Aristotle habitually does this in his Organon and all writers [B 4/8] on technical Logic follow his example. The relations connecting the things about which we reason, whether our perception of those relations depends upon the faculty of Apprehension or of Judgment or upon any other, we are upon the same ground permitted to express either by symbols definite in their interpretation or by collocation or by the union of both means. Beyond the mere substitution of symbols for words there is nothing here which is not exemplified in every living or extinct form of human speech. And hence the permission which is claimed is but the further carrying out of a principle of which the earlier writers on Logic availed themselves. Now let the system of representation thus adopted be complete and thorough-going and the processes of reasoning assumes the character of a Grammar or a Calculus. The essential laws of thought become transformed into the laws of a symbolical language and the form and value of the possible science are determined accordingly. [A11(e)/9]

[Fragment 5]

I deem it worthy of especial notice that the development $(w = A + 0B + \frac{1}{0}C + \frac{0}{0}D)^{11}$ is derived from those primary laws of thought to which attention has been directed in the earlier sections of this paper – not from a consideration of the possible relations of things or events to each other. And yet if it were our purpose to determine the possible species of such relations we should arrive at no others than are involved in the interpretation of the coefficients of that development. Let us consider for instance the relations among events to which these interpretations lead. Now the events denoted by A, B, C, D, considered with reference to the event w are represented as either 1st causally dependent upon that event to the exclusion of all other causes or 2ndly as incompatible with it, or 3rdly as flowing from it but not to the exclusion of other causes or 4thly as actually determined to exist or not to exist in perfect independence of that event.* It is easy to see in this classification the relations which have been denoted by the terms "necessary, possible, actual" as also to see how the idea of causality is modified by accompanying circumstances. Now considerations such as these do not in the Logic of Kant and [A11(h)/3] throughout the Kantian metaphysics occupy a primary place. They constitute the logical categories of the understanding. And it is a remarkable circumstance, and I would commend it to the attention of all who feel an interest in the study of the connexion between logic and metaphysics, that these categories present themselves in the present system not as primary but as secondary elements, not as requiring to be investigated on their own ground but as fully determined by the form of a mathematical development itself derived from the very laws in accordance with which alone thought is possible. [A11(f)/4]

---

* The event D is determined not to exist in absolute independence of w. But as the events $A, B, C, \& D$ include all possible contingencies this renders it necessary that some one of the events A, B, C, should be determined to exist in absolute independence of w.[12]

[Fragment 6]

The possibility of reducing the Science of Logic in its technical development under the dominion of mathematical forms, though generally perhaps regarded with distrust has nevertheless been a frequent subject of speculation. And there are few writers on the philosophy of the intellectual powers at least in recent times who have not felt called upon to express some opinion upon the question. The comprehensive mind of Leibnitz seems to have been especially possessed with the idea. He has referred to it in various portions of his writings and has stated in language which to those imbued with later and juster views of the functions of Logic must appear extravagant, his convictions of its practical importance. Some of his own speculations on the subject first published in Erdmann's edition of his philosophical writings (Berlin 1840)[13] which appear to have escaped the notice of subsequent writers are remarkable and approach more nearly to the realization of the idea of a logic de- [A7.1] veloped in mathematical forms than any other portions of his writings. To the researcher of Euler and Drobitsch, Plouquet and Lambert and in our own country of Professor de Morgan it is the less necessary here to refer as they have been made the subject of recent comment or controversy. Whatever may be thought of the special value of the above or of any similar attempts, the fact that such attempts have been made, by men independent in their habits of thought, competently acquainted with one or both of the subjects which they thus sought to bring into closer harmony and some of them eminent in general philosophy is not without significance. It creates a presumption that this consent is not arbitrary or accidental but that there exists either in the constitution of the human faculties or in the nature of things an adequate ground for the speculation.

There are indeed some plausible objections against the doctrine that the Science of Logic admits of a mathematical development or that it can properly be termed a mathematical science. For it is obvious to remark and the remark has indeed often been made that Logic is conversant with things under no limitation of kind, while mathematics is conversant with things only as they fall under the abstractions of number and magnitude. The answer to this objection will [A7.2] almost entirely depend upon what we conceive mathematics in its essential nature to be. If its subject matter can in strict propriety be limited to the conception of Number and magnitude and its method to an application of the rules of syllogism, the objection must be admitted to be perfectly valid. Our conceptions however of what constituted a Science are like almost all other human ideas subject to growth and development. The conception for instance of Astronomy must in most educated minds have undergone a deep and fundamental change during the lifetime of Newton, and the conception of chemistry a change scarcely less radical in the period of transition from the era of alchemists to that of Davy and Berzelius. Perhaps it is not too much to say that the elements of a similar change of view have gradually been evolved in connexion with the study of mathematics. The conceptions of Number and magnitude have become less prominently characteristic than the laws of thought to which these conceptions are subject – laws which certainly overstep the rules of technical Logic. Now if we consider the Science of Mathematics as no longer defined in its essential character by the nature of its subject matter but by the forms and the method of its procedure (those forms and that method having as has been said their origin in the laws of thought) the question whether Logic is or is not a mathe- [A7.3]

[Fragment 7]

A. The laws of combinations are relations among the forms which are possible *to the human mind*.

These forms are dependent upon the ideas of distribution, succession & inversion.

All the forms which are intelligible when the interpretation of the symbols is fixed are included in the forms of the mind – but not all the forms which the mind can employ are intelligible for a given set of symbols. [104]

The interpretation of the symbols determines what are the laws of combination – and also which forms are excluded, as unintelligible.

In ordinary reasoning we employ only those forms which are allowable on both grounds.

In symbolic reasoning we exclude those which are inadmissible on the former ground until we have got a symbolic conclusion when we exclude those which are inadmissible on the latter.

Operations are really only connecting links between different lawful forms just as power connects effect with cause. [C1.105]

The whole theory of symbols is included in the following propositions.

1st If a symbolic equation is rightly interpreted, if when interpreted it expresses a proposition and that proposition is true, then the symbolic [equation] is correct.

2nd. If a symbolic equation is rightly interpreted – if when so interpreted it expresses a proposition & if this proposition is false – then the symbolic equation is not correct.

3rd. If the interpretation holds, the laws of combination hold.

4th. If a symbolic premiss is correct and the laws of combination hold then the symbolic conclusion is correct.

5th. If a proposition is true then a true proposition deduced from it and from the necessary laws of things is a conclusion.

Here by interpretation is meant the putting for each symbol of its meaning. By an equation being correct is meant its being a lawful consequence of some previous propositions. Thus a symbolic premiss is correct when the proposition it represents being true and the interpretation being fixed the given equation results. A symbolic conclusion is correct when it follows from a symbolic premiss and the laws of combination. [C1.106]

Chapter XVII

Letters to Cayley, Lubbock and Penrose

————————————[1847, 1849 and 1855]————————————

Arthur Cayley

Cayley to Boole, 2 December 1847

I was very much obliged to you for the Mathematical Analysis of Logic[1] which I received a day or two ago. I have not, as you may imagine, in a subject so new to me, been able to master much of it yet. Is not the proposition that all the operations of common algebra are applicable to your symbols, p. 18, stated too generally:[2] division does not appear to me to be so; xa = xb does not imply a = b at least as I understand your definitions.

There seems to me a prima facie objection not to the principles so much as to the developability of the system: Your symbols are not combinable with numerical quantities or at any rate such combinations are uninterpretable. Has $\frac{1}{2}x$ any meaning, and if so, how does it come to be so.

I sent to Liouville a demonstration of the formulæ you sent me some time ago for some very general definite integrals, the proof was so very much the same as the one you give in the Irish Transactions,[3] that I did not think it worthwhile troubling you with it. [E 13.30]

Boole to Cayley, 6 December 1847

I am glad to hear your objections because it gives me an opportunity of replying to them. When I speak of the operations of common algebra being applicable to my system, I mean of course the symbolical operations – those which depend upon laws of combination, not upon interpretation. Thus if $(z^2-a^2)u = 0$ and z and a follow the laws of quantity, we have $(z+a)u = (z-a)^{-1}0$.

If z and a are symbols of quantity the second member is 0 but not unless. If z were $\frac{d}{dx}$ it would give a term involving an arbitrary constant – but so far as operations alone are concerned we proceed with reference only to the laws of combination. The equation $x(u-v) = 0$ gives $u-v = x^{-1}0$ whether the symbols are quantitative or elective but in the former case the result is equivalent to $u-v = 0$ — in the latter to $u-v = w(1-x)$, w being an arbitrary elective symbol. In $(75)^4$ you will find the general solution of any equation $\varphi(xyz) = 0$.

When you ask what is the interpretation of $\frac{x}{2}$ you forget that in my system x is not a quantity but represents a mental operation. All that we can desire of a calculus of logic is that it should enable us to express any *proposition* and to interpret any *equation* and to represent any [E 13.31] *act of reasoning*. The equation $\frac{x}{2} = 0$ is the same as $x = 0$ (see what I have said on the moduli in the chapter on Elective Equations[5]) and indicates the proposition there are no Xs. To ask how it is that $\frac{x}{2} + \frac{x}{2} = x$ in my system is the same as to ask how $\sqrt{-1} \times \sqrt{-1} = -1$ in a system of pure quantity for although you may interpret $\sqrt{-1}$ in geometry you cannot in arithmetic.

I think that common quantity (0×1 excepted) had nearly the same relation to my system that imaginaries have to the system of pure quantity but it is not at all necessary to enter into any such questions. They have nothing to do with the validity of the applications which I have made. [E 13.32]

Cayley to Boole, 7 December 1847

I entirely retract my objection about *division*: it is worth noticing however (I did not give it as an objection which of course it is not) that division is quite useless, from $xa = xb$ one deduces as you say $a = b+w(1-x)$, w however *is not arbitrary* but equal to $(a-b)$ and then $a = b+(a-b)(1-x)$ which is derivable from $xa = xb$ by addition and subtraction simply. I hardly see how you can say that in asking for the interpretation of $\frac{x}{2}$ I forget that x is not a quantity, my objection was that 2 is a quantity and I do not understand the combination, nor is it an answer to tell me what $\frac{x}{2} = 0$ means. The analogous difficulty in mathematics would be what does $\frac{x}{2}\frac{d}{dx}$ mean but that there is no difficulty about. You say to ask how $\frac{x}{2} + \frac{x}{2} = x$ in your system, is to ask how $\sqrt{-1} \cdot \sqrt{-1} = -1$ in a system of pure quantity. [E 13.36] To point out the want of analogy (to me) of the two questions would lead me far away into my fundamental views of analysis which I believe would be found materially to differ from yours but perhaps you have the advantage of being of the orthodox system. I wonder we should never have stumbled in our previous correspondence[6] on the subject of my *utter disbelief* of the received "English" theory of the geometrical interpretation of $\sqrt{-1}$. I would much more easily admit witchcraft on the philosophers' stone. My own theory as far as I can express it, is that a distance x, *whether real or imaginary* is an "ὄντως ὄν"[7] capable of being *measured* (that word won't do, I admit, but capable of existing) in any direction real or imaginary: somewhat as if beings whose space was of two dimensions only (which I think is conceivable) had by their science of geometry arrived at the notion of a third dimension – which e.g. the theory of symmetrically equal triangles might give them $\Delta\,\Delta$[8] an essential remark however in geometry is that the real and imaginary values of a variable x are to be considered as forming a *series*, and one is not to say that since $x = \alpha + \beta\sqrt{-1}$ that therefore the values of x form a *double series*. In fact I should admit no distinction between real and imaginary, it is only when you draw the figures that the difficulty arises. This is rather idiosyncratical [E 13.39] I am afraid, and I do not expect or wish to convert you. [E 13.40][9]

Boole to Cayley, 8 December 1847

I believe that you will find upon reexamination that the solution which I sent you yesterday is perfectly correct. The equation $xu = xv$ gives $u - v = x^{-1}$ $0 = w(1-x)$, w being arbitrary. I have only supposed x to be an elective symbol here, u and v not being necessarily so, but involving, if you please, numerical constants also. This is the most general solution. If we suppose u and v to be both elective symbols, the value of [E 13.32] u will be best got from the general theorem (75).[10] It is $u = vx + wv(1-x) + w'(1-v)(1-x)$, w and w' being arbitrary electives. This is equivalent to

$$u = v + \varphi(v)(1-x)^{11}$$

$\varphi(v)$ being arbitrary, except as respects the condition $\varphi(v)^n = \varphi(v)$. I can show that these forms are deducible from the one I sent before. They are interpretable in logic.

I must still contend that you do not give a quantitative interpretation to $\sqrt{-1}$. You have no right to introduce the word *direction* or any such word in pure arithmetic.

As to $\frac{x}{2}$ it is quite sufficient to say that its interpretation is not required. *It never occurs except in an equation* and *all equations* are interpretable in logic. I wish you would just consider this question. Can anything more be required than the *expression of any proposition the interpretation of any equation and the derivation of any results that exist*?

Try any of the solutions I have given by substitution in the original equations and you will see that the symbols which I call arbitrary are as truly so as anything in the Diff[erential] Calc[ulus].

I hope, now you have set to work to examine my principles, you will not stop short but prove them to the bottom. I do not fear the result. I had rather have one such reader as you than a thousand who take everything for granted. [E 13.33]

Boole to Cayley, 10 December 1847

Since I wrote to you it has appeared to me that I might state my principles somewhat more clearly than I have done. As respects the solution of the equation $xu = xv$ the arbitrariness is owing to the index law if $u = v + w(1-x)$ then on substitution $xu = xv + wx\overline{1-x}$ [12] but $x(1-x) = x-x^2 = 0$ by the index law, [ir]respectively of the value of w. Probably you would see this from my last [letter] but I did not state quite in such detail. You appear to me to have overlooked the index law.

And now for the numerical constants. They never appear in the expression of a proposition and all the details can be accomplished without them. Instead of taking 1 for the universe I might take u – but 1 answers quite as well and is more simple. The equation All Ys are Xs is $y = vx$ or on multiplying by $1-x$, $y(1-x) = 0$ of which the former is a solution and here and in every similar case both of expression and operation, there is nothing that cannot be interpreted. This remark applies to all that I have said in the Chapters on Conversion, Syllogism and Hypotheticals.

In the chapter on the General Properties of Elective Functions I proposed a different object viz. what are the properties of a function $\varphi(xyz)$ in which $xyz...$ are elective symbols but the function unrestricted so as to allow of its involving numerical constants? You will allow that this was a thing which I was at liberty to do. The general result is that any equation $\varphi(xyz)[= 0]$ [E 13.34] is reducible to the form

$$a_1t_1 + a_2t_2 \ ... \ + a_nt_n = 0$$

in which $a_1, a_2 ... a_n$ are numerical constants or 0 and $t_1, t_2 ... t_n$ constituent elective functions which *are interpretable in logic* – and further that this equation is resolvable into a series of equations of the form $t = 0$, *all interpretable in logic*. Prop. 2nd, p. 64f. that the numerical elements disappear altogether from the final result. I did not anticipate this, I thought it exceedingly unlikely that every equation $\varphi(xyz) = 0$ should be interpretable but when I found that this was the case and that it gave us the power of reducing the solution of equations to *general theorems* I accepted it not only as a proof that the laws I had investigated were really the laws of thought but also as a means of giving to the process of the calculus an analytical generality and simplicity which they could not otherwise have had.

I think I now might meet your objections if they still remain by this question. Do you think it necessary in order to our employing the symbol $\sqrt{-1}$ in analysis that we should be able to interpret it? Is it not in the science of quantity (apart from direction) sufficient to consider it as a symbol (i) which satisfies particular laws and especially this

$$i^2 = -1$$

and which disappears from the final result whenever a solution is real in virtue of the principle that if

$$a + bi = 0$$

and a and b are real then

$$a = 0 \qquad b = 0$$

[E 13.35] and is not this so far as it goes analogous with what I have proved in Prop. 2?[13] I contend for analogy in the nature of the things themselves and I only give this as an illustration which does not at all affect the truth of my system. [E 13.36]

Cayley to Boole, 11 December 1847

I meant to have answered your first note, and I do not think the one I received from you this morning makes much difference in what I should have said. Resuming the equation $xu = xv$ from which $u = v + w(1-x)$ — I suppose that if t be perfectly arbitrary, u and $v - t$ is a symbol coextensive in generality with w. If so, substituting for w we have

$$u = v + (u-v)(1-x) + t(1-x) \quad \text{or}$$

$$xu = xv + t(1-x)$$

but from the fundamental equation $xu = xv \therefore t(1-x) = 0$ which is a condition defining t or w ($= u-v+t$) is not perfectly arbitrary, also the equation $u = v + w(1-x)$ reduces itself to $u = v + (u-v)(1-x)$, i.e. the class u equals the class v, omitting the not-xs of v and adding the not-xs of u. A conclusion of course which follows from the original supposition of the x's of u being identical with the x's of v; but the equation

$$u = v + (u-v)(1-x)$$

is evidently derived by Addition and Subtraction merely *without division* from $xu = xv$.

You remember I did not wish to found any *objection* on this, I merely gave it as a remark – I quote for [E 13.40] explanation a passage in your first note where you say "I must still contend that you do not give a quantitative interpretation to $\sqrt{-1}$. You have no right to introduce the word direction or

any such word in pure arithmetic".[14] I think you will find in looking at my
note that I was speaking exclusively of Geometry; what I meant to say was,
that a distance x, *tho' imaginary* might be conceived to exist in any given
direction (real or imaginary) just as a *real* distance might be conceived to
exist in the same direction – that in fact the reality or imaginariness of a
distance x, has no connection whatever with the direction in which it is to be
measured – in opposition to the theory of those who maintain that the
distances a and $a\sqrt{-1}$ denote distances perpendicular to each other.
[E 13.41]

Boole to Cayley, 14 December 1847

Your argument against the perfect arbitrariness of w appears to me
quite fallacious. The direct proof that w is perfectly arbitrary rests upon this,
that if we substitute $u = v + w(1-x)$ in $xu = xv$ the equation is satisfied quite
independently of any assumption respecting w. It does not appear to me that
you meet this in any way.

I allow your substitution of $u - v + t$ for w and the result

$$u = v + (u - v + t)\,\overline{1-x} \qquad (1)$$

$$t(1-x) = 0 \qquad (2)$$

and further admit

$$u = v + (u - v)\,\overline{1-x} \qquad (3)$$

but I remark that you cannot infer from (3) that u does not involve an
arbitrary element, because the value of u is expressed in terms of u – of itself.
You ought to express the value of u in terms of other symbols before you can
assert anything respecting its elements. I remark in the second place that it
does not follow that w is not *perfectly arbitrary* because t, a part of it, is
restricted by the condition [E 13.42] (2). Two functions that are not perfectly
arbitrary may in my system produce a function that is so. The expansion of
φ(x) is

$$\varphi(1)x + \varphi(0)\,\overline{1-x}$$

neither of the terms of which is perfectly arbitrary in respect of x. Now the
coefficients of $\overline{1-x}$ in your value of u viz. (1) involves virtually both elements,
for by the solution of (2) we have

$$t = v'x$$

and the original solution from which you set out gives

$$u - v = w(1-x)$$

these together making the perfectly arbitrary coefficients of $(1-x)$ in equation (1).

This might be put in various other shapes but I hope what I have said may suffice. You have I think fallen into two misconceptions – the first is that if $u = v+w(1-x)$ and $u = v+(w-v)(1-x)$ therefore $w = w-v$; for w may and does contain an element which vanishes on multiplication into $1-x$. The second is that of inferring a property of u from an expression for u in which u *still enters*. The former misconception prevailed in your previous letter, the latter in your last. [E 13.43]

John William Lubbock

Boole to Lubbock, 16 February 1849

Let me thank you for the additional tract, which you have sent me. I have read it with great satisfaction. With such improved faculties as now exist, the province of the computer may certainly be separated from that of the mathematician, – and in carrying much further the approximations of Physical Astronomy this would, I suppose, be a point of considerable importance.

I have tried to procure two or three of your tracts advertised on the covers of those, which you first sent me, but am in- [1] formed that they are either out of print, or not known. Those for which I applied were the "Astronomical Refractions",[15] the "Numerical Coefficients" and the Essay on the "classification of the different branches of Human Knowledge",[16] the latter a subject which is to me of great interest.

Perhaps I may venture to mention to you that I have recently made a step which appears to me to be of fundamental importance in the theory of Probabilities. The ordinary (Laplace's) theory sets out from the assumption that the simple events which enter into any combination that forms the subject of experience are inde- [2] pendent and the probability of the combination an explicit function of the probabilities of these simple events. In general, however, we have no right to make any such assumption. The fact of a given combination of events having a given probability, determined by experience, is presumptive evidence that the events which are thus associated, are *not* independent.

So much as to principle – now as to method. The common theory requires that we possess just so many equations or data, as there are events, whose independence is assumed. If there are fewer data, we can do nothing unless we venture upon new assumptions [3] if there are more we are liable to incongruity and contradiction. Now the theory and the method at which I have arrived are respectively free from these objections. There is no assumption of independence except what is warranted by the logical connection of the premises and the estimation of this is a result of the *method*. There is no restriction as to the number of equations, or even as to the limitation of the elements of the conclusion whose probability is sought to the range of those which have entered into the premises i.e. into the combinations of events whose probabilities are known. When the [4] question is not determinate, the method gives the narrowest limits within which the possibility sought *must* be, and it involves (in such cases) terms which indicate the nature of that *new experience*, which is necessary to determine the probability completely. In the case in which the ordinary theory is sufficient – that in which the combination of events whose probability is sought is, so to speak, an explicit function of those whose probabilities are directly given, the result of my theory is the same as that of the former, but the method is different and, I think, superior.

Perhaps the object of my researches may be better expressed in the shape of a problem. Given the numerical probabilities of the truths of a set of propositions, these propositions being hypothetical, disjunctive or simply assertive, or combining these elements of formal distinction to any proposed extent – required the probability of the truth of *any other proposition whatever*. Or the *quæsitum*[17] of the problem may be more accurately expressed thus. Required a formulæ which shall express the numerical value of the probability sought whenever it is determinate, and which in all other cases shall exhibit [6] how much of that numerical value is determinate, and shall both indicate what new *data* are requisite to make the determination complete, and shew how to apply those *data*.

I feel that I ought to apologize for troubling you with this account. It is, I fear, too hasty a one to be quite intelligible – yet I hope that it may so far interest you as to induce you to forgive me if I have trespassed upon your time. [7]

Boole to Lubbock, 22 February 1849

Since I wrote to you I have been suffering from an attack of illness which makes it necessary for a time that I should lay aside everything [that] requires mental effort. I hope however in a week or two to be able to get to my books again and then it shall be my first business to furnish you with an abstract [2] of the theory to which I adverted in my last [letter].

The best account which I am able to give you now is by saying that a little tract on the Calculus of Logic published by me in the Cambridge Journal[18] contains principles the extent of the application of which even to Logic itself I did not at the time of its publication see – and that it is upon these principles that my theory of probabilities is based. In the first place I have found that all questions of pure Logic are reducible to the applications of one general method however numerous and however complicated [3] the premises. On applying this method to any system of propositions of which the separate probabilities are given a result is obtained which indicates the logical dependence of the proposition of which the separate probabilities are given a result is obtained which indicates the logical dependence of the proposition whose probability is sought upon those whose probability is given. The symbols which have thus far been employed are *logical*; i.e. their laws are founded upon the laws of Language – but this step being reached we are permitted in virtue of certain relations connecting the laws of thought as exhibited in [E 13.44] Language with the laws of thought exhibited in arithmetic to transform at once our single logical equation into a *system* of [4] *algebraic* equations one of which expresses the value of the probability sought in terms of a number of quantities which the other equations enable us to determine. In general the solution of these equations will depend upon that of a single central equation, the order of which is in some cases very elevated. At present I have only worked out one or two examples – and these for the elucidation of particular difficulties which have occurred to me. It is the final *logical* equation which indicates the nature of the new experience required when the probability is not quite determi- [5] nate, and in all such cases there appear terms with arbitrary coefficients in the final *algebraic* solution which our *possession* of that new experience would enable us to evaluate and so to complete the *numerical* solution.

The case of life assurance to which you refer comes under the dominion of the method.

The very simple question which I can devise as at all illustrating the character of the solutions is the following. If p be the probability of its raining at a given time, and q the probability of its both raining and [6] thundering, required the [E 13.45] probability of its thundering. For the probability sought I obtain the formula

$$q + c(1-p)$$

c being an arbitrary constant which may vary from 0 to 1. This giving for the limits of the probability sought

$$q, \ \& \ q + 1 - p.$$

It is implied by the latter that p must not be less than q as is evident from the premises. The final *logical* equation employed in the process indicates that in order to determine c we must inquire whether it ever thunders *without* rain. Suppose it ascertained that *if it does not rain* there is a probability r of its thundering, then putting r for c we have

$$q + r(1-p)$$

[7] for the probability sought. I suppose in the above case p to be probability of its raining whether it thunders or not. If we set out with the hypotheses

that rain and thunder are independent phenomena, we should get the ordinary solution $\frac{q}{p}$. [E13.46] [8]

John Penrose

Boole to Penrose, 13 March 1855

You must only consider this note as containing my hasty thoughts on the question which you propose to me.

I think that there can be no intelligible propositions about things *wholly* incomprehensible and that those writers who in their zeal to exalt the divine attributes have described them as wholly unlike what the terms by which they are expressed imply within the compass of our present consciousness or experience have by so doing only defeated their own object. The writings of the pseudo-Dionysius are the most remarkable instances of the kind which I have met with. [C9/1] But mystical writers of all ages and communions have exemplified the same form of error. Bishop Browne's Analogy to which you once directed my attention is tinctured with the same spirit. He quotes, I think with approbation, from the suppositions [in the] works above mentioned such passages as οὐκ ἀγαθός ἐστὶν ἀλλ' ὑπεράγαθος[19] etc. in reference to [1] the divine attributes.

On the other hand writers like Mill seem to me to be equally in error who maintain openly or by implication that propositions are unmeaning unless their terms relate to the distinctly conceived objects of individual experience. Though we cannot distinctly conceive of infinite space, eternal duration, perhaps of perfect goodness and purity, unchanging rectitude [C9/2] and truth etc. it does not follow that propositions in which each terms are employed are unmeaning or even that they are not positively and affirmative propositions. The solution of the difficulty is I apprehend to be found in the doctrine of the *limit*. It seems to be a law of human reason that we can in various instances affirm propositions without absolute certainty of their truth, respecting things which we can only picture or represent to ourselves as the limits of an indefinite process of abstraction. Nearly all if not all scientific truths are of this kind. We are not the less [C10/3] sure that the circle possesses such and such properties because no one ever saw and no one ever perfectly conceived by any power of imagination, a perfect circle. And so too in merely physical science. Nature never presents, and the mind I think never fully realizes by any fictive or plastic power of its aim, the conditions under which the great laws of the material universe are represented as operating in scientific Formulae. And to come to the point which you seem to have more immediately in view, the propositions which finite beings are able to express and to reason upon respecting things infinite in their nature

[C10/4] are, if true at all, true in the limit. They are propositions the terms of which we can only represent to ourselves by commencing with the finite, throwing back its boundaries and then supposing the analogous process to be continued without end. This as I have said does not appear to me [2] to render the propositions themselves less affirmative in their character. There is it is true a difference between this application of the doctrine of the limit and that which is exemplified in the propositions of ordinary science. But it is if I mistake not the [C11/5] same kind of difference as we observe between the use of the telescope and the microscope. Both will ever be *finite* in their powers, both look toward, both suggest – infinity.

You will observe that this is not the same as the question what are the grounds which we have for believing that there exists an infinite Being. I have only been considering in what light supposing such a Being to exist propositions concerning this nature and essence are to be regarded – what, in short, must be the procedure of the human mind [C11/6] in discoursing with itself of such a Being. And the purport of my observation is that as the object of all pure science are things which it would baffle all our efforts to represent by the power of imagination only and which can only be approved as the attainable limits of thought but concerning which nevertheless the most vigorous of all propositions can be *affirmed* so upon the supposition that there exists a being possessed of infinite attributes we are not precluded by the impossibility of adequately concerning those attributes from affirming respecting them clear intelligible and affirmative propositions.

And now to come to your second question. As to space and time [C12/7] I am not able to conceive of their being bounded. With respect to the Deity I find myself unable if I refer all existence back to him to conceive that he is in any way restricted as to his power or other attributes from *without*. Whatever limitations do exist must reside in the necessities of his own [3] nature. It is obvious that the subject is quite above our capacities of investigation *a priori* except as respects his moral attributes of the relations of which we may form some idea, and may, nay I think must, refer them to some *necessary* ground in himself. [C12/8]

A very deep interest attaches in my opinion to those discoveries which Science seems to be slowly unfolding respecting the plan or scheme which the Divine Mind seems to have wrought out in the work of Creation – by which term I do not mean a single set but the continued and still continuing exercise of the Creative power. I allude especially to those *Ideas* which it seems to have been the Divine purpose to realize in the constitutions of different parts of the Universe e.g. in the inorganic world, in the structure of living beings, and in the laws of the human mind. How wonderful, to take a single instance, is that [C12/9] constancy of reference to an archetypal form which is displayed in the structure of vertebrated animals from the first dawn of the existence as revealed in the ancient sandstones to the present day! Other instances will readily occur to you.

On two or three of these points I would beg to refer you to my *Laws of Thought* pp. 406-7 [and] to some passages a little further on. [C 12/10][4]

Textual Notes

Explanation of the description of the material

All texts printed here are kept in the Library of the Royal Society, London. With one exception (two letters of Boole to John William Lubbock) all texts of Boole's Nachlass are "catalogued" in the library of the Royal Society under the general signature *MS 782*. An official identification of single documents does not exist. The reason is not the lack of signatures; on the contrary each editor of the Nachlass (details in §3 of the Introduction) appears to have bequeathed his own marking system. As a rule two different signatures are found on a manuscript, as well as the paging, and references to research literature on Boole's Nachlass are likewise ambiguous. In presenting the material we have made special efforts to make it identifiable. The following elucidations attempt to show the necessity of this and to make our solution plain.

The two Signature systems "MM" and "RS"

By examining earlier editing work kept in the archives it is possible to differentiate between two different signature systems used in the material presented by us. The first is to be found in a notebook *R2* entitled "F. MacMillan, Cambridge". Read from the beginning it contains a list of signatures. The contents of the documents mentioned point these to the work of Isaac Todhunter in his rendering of Boole *1865a*. As we are not dealing with this area of the Boole Nachlass, this list does not interest us further. However, if the notebook is turned round and read from the back, there is a second list of which a separate one in handwriting with the signature *R3* exists. It is to this list we refer as the *MacMillan Signature System* or *MM-Signature* for short. Documents are listed there in telegram style, for instance as "bundle of loose papers" or, rather more helpfully, with a title or heading. The page number of the document is usually also given. No further assistance on the identification of documents is available.

Alicia Boole and H. J. Falk, who prepared a new edition of *LT*, not only sorted the material (see again Introduction §3), but also earned gratitude by documenting the various sections. We refer to these "notes upon the re-arrangement of the manuscripts of the late Professor George Boole", filed with typescripts in *MS 782*, as *Royal Society Signature System* or *RS-Signature* for short. These Notes lack any information as to contents or scope. They only record the parcel a document was in when handed over in 1889 and the parcel in which it can now be found. A description at the beginning of the list, which covers six pages of typescript, is here reproduced in full (the division lines are the original).

The Boole/Falk List of 1896

Description of the Parcel issued by the Roy. Soc. October 17th 1889.

There were nine parcels lettered A, B, C, E, R, S, T, U, W. The papers inside each parcel were numbered in pencil consecutively e.g. A 1, A 2. Only a single letter and Number e.g. A 88, S 7 was applied to any set of Papers fastened together or to any Manuscript Book.

Upon investigation it was discovered that the majority of these papers were in great disorder. Many of the separate sheets belonged to those which had been fastened together, and many sheets which had been fastened together did not belong to each other.

After inspection and consideration it was found convenient to divide the whole of the Manuscripts into three classes.
1stly. M.S. Books bound or in paper covers.
2ndly. Unbound papers upon Mathematics.
3rdly. Unbound papers upon Logic including Mathematical Logic; and Papers upon General Subjects.

Class 1 – Consisting of legible and consecutive writings remains unchanged. One of the most valuable Documents in this Class is the bound Copy (marked S. 7) of the author's "Mathematical Analysis of Logic", which is interleaved and contains a very large number of Notes by the Author.

Class 2. The Mathematical Papers were considered by the late Mr. Todhunter, though it is very improbable that all of them were in good order or that he was able fully to consider those which were not. It is certain that he did fully consider the Papers in Parcels R, T, W, most of which were published by him in the Supplementary Volume to "Boole's Differential Equations".

The Papers of this Class have received no special treatment. They remain intact in the Parcels as Lettered by the Royal Society, those in Parcels A, B, C, E, having been separated from the Logical Papers in the same Parcels and combined (still under their original Letters) in a Parcel lettered M.

Class 3. The Logical Papers received some consideration from the late Mr. De Morgan, but as so many of them were actually stitched together in great disorder, it seems at least improbable that he had a proper opportunity of estimating them.

These and the Papers on General Subjects have been exhaustively analysed and rearranged. Nevertheless, it is quite possible that some of the still unplaced fragments might be connected by even more minute enquiry. But a number of complete and consecutive Papers and Treatises have fortunately been compiled.

The Papers of this third Class have, with very few exceptions, been removed from their original parcels and arranged in three new Parcels Lettered X, Y, Z.

Parcel X contains fragments which it has been impossible to connect, or to connect more than partially.

Parcel Y contains Papers and Treatises judged to be anterior in time to the Author's "Laws of Thought".

Parcel Z contains Papers and Treatises judged to be later than the "Laws of Thought".

Some of these Papers [in parcels Y and Z] still exhibit a few lacunæ, but they are slight.

Based on the information we can take it the archive material under *MS 782* is here described in full. According to the list at the end the collection comprises 402 papers or manuscript books. In addition there are 27 typescripts which Boole/Falk had produced in the meantime. A hundred years after Boole/Falk returned nine parcels to the Royal Society in 1896, not even all sets of papers are available and whether what is found in those that are to hand was originally where it is today is uncertain. The original organisation has been kept in nine cardboard boxes which today form the whole collection (with the exception of the manuscript books). However, amongst these boxes parts of the material have often been interchanged undocumented.

From the description by Boole/Falk we may suppose that they gave the documents signatures referring to parcel and storage position ignoring the fact that signatures were already to hand. One reason for this could have been that the two systems contained in Notebook *R2* do not tally. (These systems came into being *before* 1889; Notebook *R2* and Manuscript *R3* are mentioned in the Boole/Falk list.) So it comes about that the signature "B" once marks "21 pages marked in red ink" (this in the list in R2 not considered here) and once "On the nature of Thought, 50 pages" (in the second *R2* list, used here). Boole/Falk marked "On the nature of Thought" with signatures "W4" and "C43", as the first 48 pages apparently came to light in Parcel *W* in the fourth position but the remaining two pages in Parcel *C* in 43rd position. The fact that both letter *W* and *C* from earlier editing have already been used to mark two different documents each, shows very clearly the problems to be faced when attempts are made today to work with a single signature reference. It is not unusual that several documents (or single sheets) fit to one single signature; which, then, is meant?

Details on individual documents

It is therefore necessary to identify signatures oneself. Because precision is lacking, circumstantial evidence becomes all important but does not always suffice. Therefore we give also the dimension in centimetres and the number of pages (sheets) of each manuscript. The dimensions of the typescripts are uniform (33,0 × 20,3). We also state whether a document is a manuscript ("MS") or a Typescript ("TS"). This is necessary if only because Boole/Falk simply placed on the typescripts just the signatures shown on the basic manuscript. Notebooks are marked "NB".

From Table 1 it is possible to discern in which of the Parcels *X, Y, Z* a document is filed, which gives an indication of the dating. However, due to the uncertain identification, caution should be exercised in this respect. (As regards the *MM* signature, brackets show that the placement of the document in respect of Boole/Falk parcels *X, Y, Z* is done by means of the corresponding *RS* signature.)

The compilation should also make it easier to tell whether a certain document mentioned in research literature is contained herein and in which Chapter it might be found.

In the text the original pagination is given according to a system explained below. We mark always the *end* of a page, putting the pagination in square brackets. So "[C57.22]" shows where page 22 of Document "C57" ends. Whether "C57" refers to a typescript or a manuscript is made clear in the textual notes where the signature of the source to which reference is made is given in *italics* (e.g. "*TS RS C57*" vs. "*MS RS C57*").

A dot in the signature separates the signature proper from the pagination. For example the specification "MM J.1–J.25" refers to the Macmillan signature "J" ranging from page 1 to page 25. Whereas the MM-signature incorporates the pagination in the way described and is given as a unit in the right upper corner of every manuscript page, the RS-signature is found only once in a document on the left upper corner of the front page (but caution: a "document" could also be a single page). The combination of the RS-signature proper with the original pagination in the form "RS C42.1–C42.48" is done here for sake of uniformity.

Dating on documents in the text is substantiated in the Notes. This substantiation is in most cases all too scanty and only shows how hypothetical the dating is. Despite this and since experience shows that a

Signature or main part of it	Type of document	System of signature	Boole/Falk arrangement	Printed in or as Chapter
a	MS	MM	(X)	XIV
A7	MS	RS	X	XVI
A11	MS	RS	X	XVI
A67	MS	RS	X	XVI
A88	MS, TS	RS	Y	I
A89	MS	RS	Y	II
A90	MS, TS	RS	Y	II
A91	MS, TS	RS	Z	X
A92	MS, TS	RS	Z	VI
A94	MS, TS	RS	Z	IV
B2	MS, TS	RS	Z	V
B3	MS	RS	Y	III
B4	MS	RS	X	XVI
B6	MS	RS	X	XVI
B7	MS	RS	X	XVI
B8	MS	RS	X	XVI
B9	MS	RS	X	XVI
B11	MS	RS	X	XVI
B12	MS	RS	X	XVI
B13	MS	RS	X	XVI
B77	MS	?	—	XIII
B123	MS, TS	RS	Z	XV
B164	MS, TS	RS	X	XIV
B165	MS	RS	X	XVI
C	MS	MM	(Z)	IX
C1	NB	RS	O	XVI
C9-C12	MS, TS	?	—	XVII
C24	MS, TS	RS	Z	IX
C26	MS, TS	RS	Y	II
C42	MS	RS	Z	V
C57	MS, TS	RS	Z	XI
C59	MS, TS	RS	Z	VII
e	MS	MM	(Z)	IV
E	MS	MM	(Z)	XI
E3	MS, TS	?	—	XII
E13	NB	RS	O	XVII
F	MS	MM	(Z)	VIII
G	MS	MM	(Z)	V
h	MS	MM	(Z)	XV
H	MS	MM	(Z)	V
I	MS	MM	(Y)	I
j	MS	MM	(Z)	X
J	MS	MM	(Y)	II
k	MS	MM	(Z)	VI
M	MS	MM	(Y)	II
N	MS	MM	(Y)	III
R	MS	MM	(Y)	II
U	MS	MM	(Z)	VII
W3	MS, TS	RS	Z	VIII

Table 1: Ordering of documents (signatures) to individual chapters

chronological ordering of material facilitates a first approach, we wanted to give at least an informed supposition. On this theme see "Part 3: Remarks on dating and editing the manuscripts" on page lviii.

We also note when passages have been transcribed in Rhees *1952a* or Hesse *1952a*.

Notes on the individual chapters

Chapter I: The Nature of Logic

The text is based on *TS RS A88*, a transcription of MS RS A88 = MM I.1–I.25 (22,9 × 18,7; 25 pages). On p. 5 "1848" is called "this present year". Boole/Falk classified MS A88 as being earlier than *LT*.

On the paper ribbon which holds together the pages of the manuscript someone wrote: "publish".

1 This person is also mentioned in chapter II on p. 14.
2 Whately *1856a*. This essay "on instinct" had first appeared earlier as a newspaper article, which presumably Boole had not seen.

Chapter II: Elementary Treatise on Logic

This text consists of two parts. The basis of the first part is *MS RS A89* = MM M.1–M.37 (22,6 × 18,5; 36 pages), a transcription in neat handwriting of MS RS C26 = MM J.1–J.17 (23,0 × 19,2; 17 pages). According to a note on the first page of MS C26, the transcription was made by a person named M. Lilly. This person probably also wrote MS RS E3 (see chapter XII.) Although with *TS RS C26* there exists a transcription of MS A89 we follow the manuscript of M. Lilly which has been corrected and expanded by Boole himself. Because usually reference is made to a typescript if one exists, also the pagination of the typescript is given. On the back of the first side of MS C26 someone wrote: "probably before 1849".

The second part of the text is based on *TS RS A90* which goes back to MS A90 = MM R.1–R.22 (22,7 × 18,2; 22 pages).

Boole/Falk classified MS A89, MS A90 and MS C26 as being prior to the "Laws of Thought".

On the paper ribbon which holds together MSS A89, A90 and C26 is written: "Publish".

1 This person is also mentioned above in chapter I on p. 1.
2 This is the *dictum de omni et nullo*.
3 Boole obviously speaks about the first four forms of primary propositions enumerated here on page 19.

4 Axiomata.
5 In the manuscript this large note extends over seven pages and is separated from the
  main body of the text by a horizontal stroke. Thus in the manuscript the note does not,
  as in the typescript, separate the two sentences here printed directly before and after
  the note, i.e. "This question does not belong however either to Logic properly so called
  or to geometry" and "The successive steps in the demonstrations of geometry are
  usually syllogistic ...".
6 Book 1.
7 Newton *1707a*.
8 In the manuscript the expression "modern Algebra" is replaced by "the science".
9 In the manuscript the expression "Calculus" is replaced by "science".

Chapter III: [Extracts from a notebook]

These are the extracts from a suite of short essays mentioned in §8 as
written in a notebook and then taken out by Boole. They come from *MS RS
B3* = MM N.1 – N.30 (22,7 × 18,7; 79 pages) Boole/Falk classified this
manuscript as being prior to *LT*. Their content relates closely to "Sketch"
published by Rhees in Boole *1952a*, 141-166, which dates the text at around
1849. For discussion of this point and its context, see Panteki *1992a*,
540-552, who first drew attention to these essays.

The chosen passages, which are separated by a line of stars *, come from
respectively folios 1, 8v, 9, 10v-14, 15-16v and 17-18; the continuous line in
the sixth extract is Boole's own. The rest of the manuscript up to N.30
includes essays or notes on primary and secondary propositions, and on elimination.

1 Boole is referring to his three basic algebraic laws (6.1-3).
2 Dirichlet *1839a*.
3 After writing and modifying these notes, Boole imposed a numbering by clauses onto
  them which partly overrode some original numbering and somewhat changed the order
  of material. We follow this latter order, and eliminate the old numberings which seem
  to have been replaced. Despite this housework, the text is still a little untidy.
4 Four unreadable words here.

Chapter IV: Prolegomena

The text is based on *TS RS A94*, a transcription of MS RS A94 = MM
e.1– e.20, numbered backwards (first four pages 32,5 × 20,1, remaining pages
33,5 × 20,6; 20 pages in all). The unusual dimensions and other
characteristics of the manuscript are very similar to the "Preparatory Notes"
(chapter IV). These were classified by Rhees as preparations for
"Foundations of the Mathematical Theory of Logic" (chapter V). Judging
solely by the appearance of the manuscript, the "Prolegomena" could
therefore also be preparatory to this work. The first editors of the Nachlass
were evidently also of this opinion; compare note 12 to chapter V.

Boole/Falk classified MS A94 as being later than *LT*. This is justified by the fact that Boole makes a reference to this work on p. 61.

Hesse *1952a* takes a citation out of this text.

1 In a first attempt Boole described "except" as a conjunction, then he changed it into "preposition".
2 The typescript corrects a mistake of Boole. He used here the property "chew the cud" instead of "divide the hoof".

Chapter V: On the Foundations of Logic

This text consists of two parts. The first part is based on *TS RS B2*, a transcription of MS RS B2 = MM G.1–G.39 (23,0 × 19,0; 48 pages), in neat handwriting. The second part of the text is based on *MS RS C42* = MM H.1–H.46 (23,0 × 19,0; 47 pages), neat handwriting. The originals for MSS B2 and C42 – a bundle of around 60 pages in various sizes, some of them stuck together – are also to be found in the archives.

LT is mentioned in the first sentence of MS B2 as being published "about two years ago". On the back of MS C42 Mary Boole wrote "later than 1855" (the year of her marriage). Both manuscripts have been classified by Boole/Falk as being later than *LT*.

Rhees published some extracts in *1952a*, 230-246 as chapter vii and some of the material he used in his "Note on Editing" of the same work. This comes to about one third of the whole. Hesse *1952a* cites seven short passages which amount to at most a tenth of the text.

1 Boole *1857a*.
2 Charles Hughes Terrot (1790-1872) studied at Trinity College Cambridge, where he was an associate of William Whewell and George Peacock. A good mathematician, he was a Fellow of the Royal Society of Edinburgh, contributing many papers. On the point mentioned by Boole here, he expressed himself in Terrot 1857a.
3 Boole *1857a*.
4 "Hê logikê", means "the logic".
5 "Hypostasis", here in the meaning of "subsistence". The stoic difference between existence and subsistence referred to here by Boole, is to be understood as analogous to the theory of Alexius Meinong. Here Boole muses on "the notion in the concrete individual *existence*" as opposed to the notion in the condition of subsistence. The "subjective unity" of which Boole speaks in this connection is the unity of the subject and has nothing to do with subjectivity in its modern sense.
6 "Noêton", means an "intelligible".
7 "Aisthêton", means a "sensible". The sentence reads in English: "... it is an intelligible rather than a sensible."
8 In this footnote Boole first alludes to Hamilton's famous review *1833a* on Whately; in the reprint which Boole probably read, see *1852a*, 137n and 158-159. His citation of Suarez's *Metaphysica* (1605a, vol. 2, 516-518) on second intentions is a remarkable example of his willingness to go back to old sources and moreover on a topic which was

less well understood in Boole's time than in Suarez's. An edition of Suarez's *Opera omnia* was in progress in Boole's lifetime, but *Metaphysica* came out only in 1877, as volume 25.

9 It is not clear to which edition of Whately *1826a* Boole refers here. In the first edition a possible reference is found on p. 88. See Introduction §5 for a comment on this passage.

10 Hamilton *1833a*, 134.

11 The original of MS B2 has here room only for a later reference, no brackets. The first editors of the Nachlass completed the brackets in their transcription with the reference "A7". The first three pages of this manuscript are reproduced here as Fragment 6 of Chapter XVI.

12 The phrase "introductory portion of this paper" has been changed in transcription in a reference to a "Cp. A.94". This makes Chapter IV "Prolegomena" (= RS A94) an introduction to the present Chapter V. The theme "Abstraction" is dealt with in Chapter IV in page 58.

13 R. G. Latham (1812-1888), ethnologist and philologist, wrote a volume *1847a* on the *First notions of Logic* which Boole praised in footnotes in *MAL*, 5 and *LT*, 40 but criticised in *MAL*, 59. Boole wrote to him on 22 May 1855 on the forms of plurals in language (copy at MS RS C 7-8; transcribed in Panteki *1992a*, 544).

14 Boole probably had in mind a passage in De Morgan such as *1847a*, 67-69.

15 This seems to indicate a reference to some other work by its section number. The obvious candidate is *LT*: if so, the appropriate sections would be 7 –11.

16 Reference in empty brackets supplied according to Boole *1952a*.

17 The number of the equations (1) and (2) and all later cross-reference by this number to this equation is supplemented by Boole *1952a*.

18 Between the pages originally numbered 23 and 26 there is only a single page numbered 24(+25). The pagination incorporated in the MM-signature is false too: between page 11 and 13 there are two pages 12 and 12'. Because of these two faults, the last page of the manuscript is in one counted p. 48 and in the other p. 46, the exact number of pages being 47.

19 Addition of the equation in Boole *1952a*, 243. Originally there was only a pair of empty brackets '( )'.

20 This is written on a small piece of paper stuck on the back of C42.36. There is no reference explicit or implicit, to the texts of C42. 36 or C42.37. Perhaps Boole stuck this note wrongly on the back of C42.36 instead of C42.37 where it would be directly opposite to the text here.

Chapter VI: [Preparatory Notes]

The text is based on *TS RS A92*, a transcription of MS RS A92 = MM k.1 – k.9 (33,5 × 21,1; 9 pages). Boole/Falk classified this manuscript as being later than *LT*. This work is mentioned by Boole twice on p. 110.

According to a remark on the last page of the manuscript this last page was probably originally the first page and then sewn up wrongly. We therefore bring p. 11 first.

The title given comes from Rush Rhees. In Boole *1952a*, 246 Rhees uses a short citation from what he declares to be "preparatory notes" for "On the Mathematical Theory of Logic" which is printed here as chapter Chapter V.

If Rhees is correct, then MS A92 must be earlier than MS B2 and MS C42, written in 1856.

The division lines in the text are the original.

1 A note on the top of this page says: "This page sewn last in MS is probably the first."
2 This is the first sentence on the top of p. 1 of MS A82, placed before the title "Definitions of Conception, Judgment and Reasoning".
3 In MS A92 the paragraph "Analysis of conceptions" which follows here on the next page faces this passage. The paragraph is written on a separate sheet and stuck on the verso of the preceding page. It seems that it should be read as an addendum to this passage.
4 Compare note 3.
5 In the manuscript there is at this place a big gap of 12 lines, probably for the insertion of a title.
6 This and the next two paragraphs up to "... among the different sciences." are in MS A92 found on the back of the preceding page. Compare note 3.

Chapter VII: General Summary

The text is based on *TS RS C59*, a transcription of MS RS C59–C65 = MM U.1–U.12 (23,2 × 18,6; 12 +1 pages). The manuscript is a complete fascicle, the title is written on the frontpage, which contains no other text; and the last page is empty.

Boole/Falk classified this manuscript as being later than *LT*. This work is mentioned on pp. 113 and 115.

Chapter VIII: [Preface]

The text is based on *TS RS W3*, a transcription of MS RS W3 = MM F.1–F.5 (23,0 × 18,8; 5 pages), neat handwriting. Boole/Falk classified this manuscript as being later than *LT*. This work is mentioned on p. 119.

Hesse *1952a* takes two short citations from the preface. She refers to the work simply as "a sequel to *LT*".

The division lines in the text are the original.

As a part of MS RS C27 = MM b.3–b.43 there exists another "Preface" for a sequel of *LT* in the archives. It covers the first four pages of the manuscript C27, then the text was struck out.

1 This note is written in pencil by Boole at the top of MS W3.1 without a reference into the text. "Mr. M." is presumably MacMillan, publisher of *LT*. It could be Daniel MacMillan (died 1857) or his younger brother Alexander.
2 This paragraph (written in pencil) comes after the end of the main text (written in ink), separated from it by a line.

Chapter IX: Table of Contents

The text is based on *TS RS C24*, a transcription of MS RS C24 = MM C.1–C.5 (22,9 × 18,1; 5 pages). Boole/Falk classified this manuscript as being later than *LT*. This work is mentioned on p. 122.

Hesse *1952a* gives the titles and a condensation of the description of the content. She thinks that the text is part of the planned sequel to *LT*.

1 Quite likely Mansel *1860a*. (Compare his praise in the manuscript published in Boole *1952a*, 212.) This note is found on the back of the (empty) front page of MS A91.

2 The word is hardly legible. In TS C24 is written "present" and later added in square brackets "parent?".

Chapter X: The Philosophy of Reasoning

This text is based on *TS RS A91*, a transcription of MS RS A91 = MM j.1–j.4 (32,5 × 20,5; 4 pages). On the back of the single sheets are notes on geometrical problems. Boole/Falk classified this manuscript as being later than *LT*. This work is mentioned twice on pp. 123f. On p. 124 *LT* is called a "predecessor" of the planned work.

Hesse *1952a* cites a passage from this work which is described by her as "unfinished paper which may have been intended to be the introductory chapter."

There is another manuscript in the archives which bears the title "The Philosophy of Reasoning". It bears the signature MS RS E10, also marked A1–A11.

1 Hamilton *1833a*, Whately *1826a*.

2 The second half of this paragraph is our construction from a messy and altered text. Left loose is the clause "permits such glimpses only as serve to awaken curiosity". If its implied subject is "introspection", as the context suggests, then we have before us here an important statement of scepticism by Boole.

Chapter XI: Logic

The text is based on *TS RS C57*, a transcription of MS RS C57 = MM E.1–E.52 (28 × 19; 54 pages of varying sizes, partly stuck together). On the back of page 16 of the manuscript is written "Later than 1855", which is probably a note of Mary Boole. In any case Boole/Falk classified this manuscript as being later than *LT*.

Concerning the dating of the manuscript Daniel D. Merrill points out (private communication) that a controversy between Boole and De Morgan which appears here on page 140 seems also to reappear in the correspondence between them. In a letter dated 13 July 1860 De Morgan replies to Boole: "I agree with you that the explanation of ($\cdot$) and) (is not down at the bottom." This might echo Boole's point that De Morgan's interpretation of these symbols comes from his rules of inference and not from the inherent meaning of his symbolism. The corresponding letter of Boole is not in Smith *1982a* so this assumption cannot be checked.

Boole's representation of De Morgan's symbolism makes the impression that he had oriented himself to his article on "Logic" in the English Cyclopedia (*1860a*, 260f.). In the letter mentioned above De Morgan refers himself to his *1860a*. Boole replied by return, that "I have not seen and fear that I shall not see here the English Cyclopedia", but "if ever I do get access to the work I will not fail to turn first to your paper" (Smith *1982a*, 80). If De Morgan sent him an extra copy thereafter is not known. But it is interesting that in the formulation of De Morgan's second rule of notation (page 139) Boole uses the opposite pair "total / partial", as Merrill points out. In De Morgan *1850a*, 35-37 – another possible source for Boole – the pair "universal / particular" is used instead.

We cannot represent De Morgan's system here but we give passages of his *1850a* and *1860a* where he made statements to which Boole may have referred.

In Boole *1952a* and Hesse *1952a* there are short citations which amount to 17% of the whole text. Hesse calls MS C57 a "draft of a chapter which seems to be almost complete and which covers the ground of I in the Table of Contents. It is therefore almost certainly intended to be the first chapter of the book."

1 "Logos", means "word, thought, reason".
2 This is obviously a reference to the first article of the paper. As the further references show, where the number of the article is not yet inserted, Boole obviously planned to number the paragraphs later.
3 Boole cites the first edition of Hamilton *1852a* from a section of "Appendix II. Logical" which defends the form "Some Xs are not some Ys" against criticism by De Morgan. The passage cited reads (*1852a*, 640*, *1853a*, 695):
   Whatever is operative in thought, must be taken into account, and consequently be overtly expressible in logic; for logic must be, as to be it professes, an unexclusive reflex of thought, not merely an arbitrary selection – a series of elegant extracts, out of the forms of thinking.
4 Baynes *1850a*, 74; Boole is quoting him, from "a negative" to the end of the sentence.
5 In the original and its transcription the second rule of notation comes before the first one. This shows the hasty character of the manuscript. The person who made the transcription was obviously not acquainted with De Morgan's notation. The obvious corrections have here been made without remark.
6 De Morgan *1860a*, 261. Compare De Morgan *1850a*, 31: "Let the inclosing parenthesis, as in X) or $(X$, denote that the name-symbol X, which would be inclosed if the oval were completed, enters universally. Let an excluding parenthesis, as in)Y or $X($, signify that the name-symbol enters particularly."
7 De Morgan *1860a*, 261.
8 Compare De Morgan *1850a*, 35f.

9 De Morgan *1850a*, 31: "... the canon of the inference, when there is one, is,—*Erase the symbols of the middle term, the remaining symbols shew the inference.*"

10 De Morgan *1850a*, 35f. and *1860a*, 123.

11 Daniel D. Merrill defends De Morgan against Boole's reproach that the symbols are not interpreted by a rule "deduced from the interpretation of the symbols". Merrill says (private communication): "In fact, De Morgan provides just such an interpretation. Thus, the rule is deduced from *De Morgan's* interpretation of the symbols. In *1850a*, 36-37 De Morgan provides technical definitions of the 'universal/particular' and 'affirmation/negation" distinctions, as he does in the *1860a* article, where the former distinction becomes the 'total/partial' distinction (*1860a*, 260-261)." Starting from these characterizations Merrill shows that "De Morgan has a systematic theory from which these interpretations follow. This theory is prior to his rule of transformation, let alone his syllogistic cancellation rule."

12 In the original there are two variants of this sentence melted into one and the sentence ends with "... the error of a too narrow definition is seen."

13 We read "connecting" in MS C57, TS C57 has "converting".

14 The text breaks off here. The last word is illegible.

Chapter XII: On Belief in Its Relation to the Understanding

The text is based on *TS E3*, a transcription of MS E3 (22,8 × 18,4; 14 pages), in neat handwriting. The transcriber was probably the same person who wrote out MS RS A89 (i.e. the first part of chapter II "Elementary Treatise on Logic")

Boole/Falk classified a manuscript "E3" as fragmentary or impossible to connect to others more than partially. But the manuscript here seems to be complete (compared with others). Therefore we do not count the signature "E3" as the RS signature.

The manuscript has a note on it written in pencil: "Should be published". At the back of the manuscript is written: "Belief in relation to the understanding, New forms of Syllogisms, De Morgan etc." Because syllogisms and De Morgan are not mentioned in the present text we could here have a fragment of a larger work (or the manuscript was bundled together with others).

Chapter XIII: The Philosophical Idea of Freedom

This text is based on *MS B77* (23,0 × 19,2, 3 pages), not paginated. If the first sentence on the frontpage [B 77.1] could be read as a headnote, as we have assumed, then the text should be complete and not fragmentary.

It is uncertain if the signature corresponds to the Boole/Falk arrangement because according to this list "B 77" has been classified as a mathematical paper. But the present text does not contain any formulæ at all. "B 77" is therefore not an RS-signature.

The division lines in the text are the original.

1 "Cosmopœa" (or "Cosmopœia") is Greek for "creation".

Chapter XIV: Note [to Aristotle]

This text is based on *TS RS B 164*, a transcription of MS RS B 164 = MM a.1 – a.6 (23,7 × 19,0; 6 pages). Boole/Falk classified this manuscript as fragmentary or impossible to connect to others more than partially.

There is no indication of the date of creation. Boole *1854a*, 49 cites exactly this passage, but the context is different.

1 Old numbering. The book Γ (= Gamma, i.e. the third letter of the alphabet), from which the quotation comes, is today called the fourth book. The reason for this is that Book A (= Alpha, i.e. the first letter of the alphabet) counts today as two books.
2 Boole notes as textual variant: "The inquiry concerning these *axioms* also would belong to one who considers ..."
3 "And physics also are a kind of philosophy, but not the first." The remark was made by Boole on the verso of the foregoing page without direct reference into the text, but situated at the same height as the text to which the note is here attached.
4 Addition made by Boole.

Chapter XV: Philosophy of Mathematics

The text is based on *TS RS B123*, a transcription of MS RS B123 = MM h.1–h.37 (17,0 × 20,5; 37 pages). Boole/Falk classified MS B123 as being later than *LT*. The division lines in the text are the original. Hesse *1952a* takes four short citations out of this text.

1 Barrow *1683a*.
2 Boole quotes here Hallam *1829a*, 369. We have rendered the text exactly as in the original; Boole was accurate apart from a few details of punctuation. The exception is "circle (sphere)", where the second word is Boole's correction of Hallam's slip in conception. Boole cited p. 309 of this volume in *MAL*, 33 over an issue concerning the history of syllogisms. Henry Hallam (1777-1859) was a well-known cultural historian whose writings certainly deserve revival.
3 Eutheia grammê estin, hêtis ex isou tois eph' heautês sêmeiois keitai. This is the original Greek of Boole's citation "A straight line ...".
4 Gap in the manuscript (between the last word on h.30 and the first word on h.31).
5 Ex isou, means "evenly".
6 Kai ta epharmozonta ep' allêla isa allêlois estin. This is the original Greek of Boole's citation "Magnitudes which coincide ..."
7 Euclid *Elements*.
8 Aitêmata.
9 Koinai ennoiai.

Chapter XVI: [Various fragments, apparently late]

The seven fragments united in this chapter are taken from three different works or sources. We discuss each work or source separately.

The basis of the first five fragments of this text are a selection of five bundles of manuscripts ($22,0 \times 18,2$; $20 = 2+7+4+5+2$ pages) in the same neat handwriting which also distinguishes other papers. They seem to belong to a single, fairly advanced work. Boole himself seems to have scattered the manuscript – this would give an explanation as to why the signatures on the single pages do not give a coherent picture. This is true for the signatures on the top left side and the independent pagination on the top right side none of which has a logical order.

To show this we give here an overview and note the signature and the pagination of a page as a pair divided by "/". Fragment 1: *A11(b)*/1; *B7*/2. Fragment 2: *A67*/1; *A11*/2; *B8*/3; *B6*/1; *A11o*/2; *A11(k)*/3; *A11(g)*/4. Fragment 3: *B11*/5; *A11n*/2; *B165*/[illegible]; *B9*/4. Fragment 4: *B13*/5; *B12*/6; *A11(d)*/7; *B4*/8; *A11(e)*/9. Fragment 5: *A11(h)*/3; *A11(f)*/4. It seems that the extensive Boole/Falk re-arrangement came here to its limits. We count the signatures "A11(b)" etc. as RS signatures with some reservation, supposing that the small Latin letters "(b)" etc. were later additions.

Probably Boole split his work into parts in a new attempt to write a successor of *LT*. For example, the last two lines of fragment 3 have been cut off and may lie somewhere else. Maybe a part of the whole work ended as scrap paper. The result is that most fragments are too small to have independent value. We selected some which are relatively complete. In fragment 1 we only suppressed the last paragraph and left out the first in fragment 4.

In three of the manuscripts reference is made to *LT*; on fragment 2 Mary Boole wrote "Later than 1855". Only fragment 5 is without textual evidence of the date of its origin.

Hesse *1952a* has citations from fragment 1 and 4.

Fragment 6 is taken from *MS RS A7*, a bundle of 12 pages with varying dimensions ($23,0 \times 17,4$ to $18,6$), numbered 1–3, 5, 8, 20, 22–25, 38–39. Boole/Falk classified this manuscript as fragmentary or impossible to connect to others more than partially.

The philosophical groundwork is laid in the first pages. We bring the first three pages only. Unfortunately page 4, where Boole apparently analyses the important question "whether Logic is or is not a mathematical [science]" is missing. There must have been something relevant on it as page 5 starts with the statement: "This is the question which I propose to endeavour to cover in the present essay." On p. 20 the system of Notation is explained, pp. 22ff. deal with the conditions of interpretability and pp. 38f. consider the method of development and the name "Dual Algebra" is mentioned.

The first page of the sixth fragment bears in the left upper corner the note: "All this is written by G. Boole on the backs of paper copied by me (and no longer wanted). ... it is all later (probably much later) than our marriage in 1855. M. Boole". On the back are hasty calculations. Obviously Mary Boole found these more important than the text she once copied.

The neat handwriting of fragment 6 seems to be the same as that on the first five fragments. But the paper is different – blue instead of white – and the first five fragments have no writing on the back.

Fragment 7 was originally in a notebook bearing the signature RS C1, and also marked as "6 Logic" (23,6 × 19,0). A similar marking, by the way, is to be found on other notebooks, the signatures "2 Logic" (RS E5), "3 Logic" (RS E1) and "5 Logic" (RS C66) being in existence.

The particular one mentioned has a marble-patterned cardboard cover and is not paginated, margin notes referring to personal numbering. It contains densely-packed entries on the most widely different questions of logic. A copy of the original page can be seen above p. lxi.

Hesse *1952a*, 74f., expands on that part of Notebook 6 from which our fragment comes. For a first analysis we refer to these detailed discussions. They concern three statements therein called Boole's 'three axioms' for "symbolic methods" which are found, with others, on pp. 87-90 of the notebook. There Boole is trying to research the logical dependencies between certain statements through symbolic derivation. The letters of his logical equations represent the following: p = real premiss, p' = symbolic premiss, i = interpretation, l = laws of combination, y' symbolic conclusion, y = real conclusion. Or: p = proposed relation is true, i = symbols are rightly interpreted, s = symbolic proposition is true. Evidently Boole is dealing with the bridging between reality and symbols.

The statements looked at in this part of the notebook are neither commented upon nor reasoned out by Boole. They simply stand there as a starting point or translation of lettered equations. The theme is merely indicated with a short heading. By "Reasoning" stands: "If premises are true and if reasoning is correct the conclusions are true." [87] (= Axiom 1 in Hesse.) Under "Language" there is the proviso: "If a proposed relation among things is true, and if the symbols by which that relation is expressed are rightly interpreted, then the symbolic proposition which expresses that relation is true." [89] (= Axiom 2 in Hesse). The long calculating out that follows apparently failed to reach its target for Boole crossed it all out and began anew with: "If a symbolic proposition is true and if it is rightly interpreted then thus the relation among things holds." (= Axiom 3 in Hesse).

Paragraphs on "Parallel Reasoning in different languages" [92] and "Symbolic Reasoning" [93] follow the paragraphs "Reasoning" and "Language". After several pages stuffed with calculations fragment 7, which we print here, appears. In it Boole first concerns himself with the problem of unintelligible symbols. After this he leaves about half a page empty (here shown by a dividing line) to begin at the top of a new page with "The whole theory of symbols ...".

After fragment 7 there are around three pages of formulas, and then the entries stop.

1 The roman numerals in the references in brackets on this and the next page are supplied by the editors.

2 Reference missing.

3 Number of the chapter supplied by the editors.

4 Leibniz *1681a*.

5 "a for animal, b for rational, c for mortal, d for visible."
  In the manuscript d stands for "possibile", but in the following example Boole uses d for "visibile". For this reason the text has been adapted by the editors.

6 "Terms are a, b, ab, bcd, for example: animal, rational, animal rational, rational mortal visible. A universal affirmative proposition I designate as follows: a is b, where a is subject, b predicate. Is: copula."
  The citation of Leibniz is adapted by Boole. The full text of Leibniz is: *"Terminus est a. b. ab. bcd ut: homo, animal, animal rationale, rationale mortale visibile. Propositionem universalem affirmativam sic designo: a est b, seu (omnis) homo est animal, semper enim hic signum universalitatis intelligi volo, ubi a subjectum et b praedicatum. Est: copula."*

7 The citation of Leibniz is adapted by Boole. Leibniz wrote, "Sufficit enim dici a est b ..."

8 "2) Transposition of letters within the same term doesn't change anything, e.g. ab which coincides with ba or *animal rational* and *rational animal*. 3) Repetition of the same letter within the same term is useless, e.g. b is aa, or bb is a; man is animal animal, or man man is animal. For it is sufficient to say b is a or man is animal."

9 Addition and square brackets by Boole.

10 Reference left empty. Probably to chapter II, see esp. *LT* p. 30.

11 In the manuscript there is only one pair of brackets left empty for later reference. The formula is supplied by the editors. To reconstruct the development of which Boole may speak here compare this fragment with the specific interpretation of the coefficients of a development on p. 99 above. There Boole brings the coefficients into connection with the categories "existent, nonexistent, indefinite, impossible". In *LT* p. 92 a fully developed formula is generally given as "$w = A + 0B + \frac{0}{0}C + \frac{1}{0}D$". To fit the description which Boole gives in this fragment one has to exchange in this formula the last two coefficients and thus write the category of the indefinite at the end, as Boole did in Ch. V, p. 98 above.

12 To avoid an open contradiction with what is said about the independence of the event D in the text, the note could be read as a hypothesis leading to a *reductio ad absurdum*.

13 Leibniz *1840a*.

Chapter XVII: Letters to Cayley, Lubbock and Penrose

The original of the seven letters exchanged between *Arthur Cayley* and George Boole is a copy in neat handwriting in Notebook *RS E 13* (24,6 × 20,0), which contains also the copies of letters to Lubbock (see below). The sheet of the notebook with pages 37 (front) and 38 (back) seems to have been cut out before the text was written down; the text runs without interruption from page 36 to page 39. The letter of December 7th 1847 is wrongly placed between the letters of December 10th and 11th respectively. Boole sent his letters from Lincoln, Cayley probably from London.

Of the two letters of Boole to *John William Lubbock* we have not only the copies but also the originals in the Library of the Royal Society, London. The original letter of Feb. 16th 1849 bears the signature *LUB.B362* (18,6 × 11,2) and the original letter of Feb. 22nd 1849 the signature *LUB.B363* (18,2 × 11,2). In Notebook *E13* Boole made copies of these letters following the letters to Cayley. The copy of the second letter is not complete; we follow the selection of Boole and leave out some of the introductory remarks and the postscriptum. They concern the difficulties of obtaining books and offer solutions. The copy in NB E13 begins with "The best account which I am able to give you now ...". It was published by Jourdain using a copy in the lost Harley volume of letters (*1910a*, 335-336).

The simple pagination in the text [1], [2], etc. refers to the letters LUB.B362 and LUB.B363. The pagination [E 13.44] – [E 13.46] refers to Boole's transcription. In both cases Boole sent his letters from Lincoln.

The basis of the letter to *John Penrose* is TS C 9–12, a transcription of *MS C9-12* (19,0 × 11,6; 10 pages). The typescript has no original signature; someone has added later in pencil the signature "C 9–12". Each page of the manuscript is numbered on the right side at the top consecutively from 1 to 10 and each sheet (containing 2 to 4 pages) is numbered on the left side consecutively from C9 to C12. We give both signatures and the paginations of the manuscript divided by "/". A single number refers to the typescript C 9–12. Boole sent this letter from Cork.

Boole/Falk classified some MSS C9 – C12 as fragmentary or impossible to connect more than partially. This classification can hardly be true of the letters concerned and thus the signature does not belong to the RS system in our sense.

1 *MAL*.

2 At *MAL*, 18 only *direct* processes are mentioned. Boole says there that "... all the *direct* processes of common algebra are applicable to the present system". This leaves open the possibility to consider division as an inverse operation.

3 Cayley was referring to Boole *1848b*; our date is that of the title page of the journal, but obviously this part was out earlier. Cayley *1848a* commented in Liouville's journal on Boole's note *1848c* on this theme.

4 Prop. (75) in *MAL*, 73.

5 *MAL*, 60ff.

6 MacHale *1985a*, 55ff. gives excerpts from the Cayley–Boole correspondence in the years 1844 and 1845, now in the Library of the University of Cambridge.

7 "Ontôs on" means "a real being" (verbally "a being being").

8 The meaning of the two triangles is not clear. It is also puzzling that they do not separate two sentences: the following word starts with a lower-case letter.

9 Page 40 of notebook E13 continues with the letter from Cayley to Boole of December 11th 1847.

10 See note 4 above.

11 The manuscript has w instead of u as first term of the equation, i.e. $w = \underline{v + }\varphi(v)(1-x)$.

12 The manuscript has as first term of the equation only x, i.e. $x = xv + wx\,\overline{1-x}$.

13 *MAL*, 64.

14 Obviously Boole to Cayley on December 8th 1847, above p. 193. But it is unclear why Cayley calls this letter the *first* note of Boole

15 Lubbock *1850a*. [?]

16 Lubbock *1838a*.

17 I.e. the object of investigation.

18 Boole *1848a*.

19 "Ouk agathos estin all' hyperagathos", means verbally "he is not good but more than good". The negation in this sentence expresses in this context that he [god?] is not a possible object for the predication of "good", i.e. it is equally true that he is not not-good.

Bibliography

A full bibliography of Boole's published writings is contained in MacHale *1985a*: that in Smith *1982a* has few gaps. Regarding historical literature, this bibliography includes the most notable contributions but excludes the repetitive scribbles that all famous names attract, and also the surveys in general histories of logic, mathematics or philosophy that are often satisfactory in their contexts but do not contain original views.

If more than one edition of a work is listed, then "†" marks the one to which page numbers in the text are given. Works *cited* by Boole in the texts published in this edition are prefaced by "∗".

Anonymous *1865a*: "The Late Professor Boole", *The Canadian Journal 10*, 44-45.

∗ Aristotle: *Metaphysics*.

Auroux, S. *1981a*: "Condillac ou la vertu des signes", in Condillac, *La langue des calculs* (1798), ed. and intr. Auroux and A.-M. Chouillet, Presses Universitaires, Lille, i-xxxviii.

∗ Barrow, I. *1683a*: *Lectiones Mathematicæ*, London.

∗ Baynes, T. S. *1850a*: *An Essay on the New Analytic of Logical Forms*, Sutherland and Knox, Edinburgh.

Blakey, R. *1851a*: *Historical Sketch of Logic, From the Earliest Times to the Present Day*. Nichols, Edinburgh.

∗ Boole, G. *1835a*: *An Address on the Genius and Discoveries of Sir Isaac Newton* (Delivered on Thursday, Feb. 5, 1835), Gazette Office, Lincoln.

— *1844a*: "On a General Method in Analysis", *Philosophical Transactions of the Royal Society of London 134*, 225-282.

— *1847a* [cited as *MAL*]: *The Mathematical Analysis of Logic, Being an Essay Towards a Calculus of Deductive Reasoning*, Macmillan, Barclay & Macmillan, Cambridge and Bell, London†. Reprinted Blackwell, Oxford; and Philosophical Library, New York, 1948. Also in *1952a*, 49-124 [original pagination marked, and used here]. French transl. "Analyse Mathématique de la Logique" in: *Algèbre et logique d'après les textes originaux de G. Boole et W. S. Jevons avec les plans de la machine logique*, trans. F. Gillot, Blanchard, Paris 1962, 13-88. Italian transl. *L'analisi matematica della logica*, Boringhieri, Torino 1993.

∗ — *1848a*: "The Calculus of Logic", in *The Cambridge and Dublin Mathematical Journal 3*, 183-198. Also in *1952a*, 125-140.

— *1848b*: "On a Certain Multiple Definite Integral", in *Transactions of the Royal Irish Academy 21*, 140-149.

— *1848c*: "Théorème général concernant l'intégration définie", *Journal des mathématiques pures et appliquées* (1) *13*, 111-112.

— *1848d*: "MM. [sic] Boole's Theory of the Mathematical Basis of Logic", *Mechanics Magazine 49*, 254-255.

— *1851a*: "On the Theory of Probabilities, And in Particular on Mitchell's Problem of the Distribution of Fixed Stars", *Philosophical Magazine* (4) *1*, 521-530. Also in *1952a*, 247-259†.

∗— *1854a* [cited as *LT*]: *An Investigation of the Laws of Thought on which are Founded the Mathematical Theories of Logic and Probabilities*, MacMillan, Cambridge and Walton & Maberly, London†. [Reprinted Dover, New York, 1958. 2nd ed., ed. P. E. B. Jourdain, Open Court, Chicago, 1916, repr. 1940 and 1952. French trans.: *Les lois de la pensée*, trans. S. B. Diagne, Vrin, Paris, 1994. Italian transl. *Indagine sulle leggi del pensiero*, Einaudi, Torino 1976.]

∗— *1857a*: "On the Application of the Theory of Probabilities to the Question of the Combination of Testimonies or Judgments", in *Transactions of the Royal Society of Edinburgh 21*, 597-652. Also in *1952a*, 308-385. (Read 1857.)

— *1859a*: *A Treatise on Differential Equations*, MacMillan, Cambridge and London. [Later editions by I. Todhunter 1865 (see Boole *1865a*), 1872, 1877†.]

— *1860a*: *A Treatise on the Calculus of Finite Differences*, MacMillan, Cambridge and London. [Later editions by J. F. Moulton, 1872 and 1880 German trans.: *Die Grundlehren der endlichen Differenzen- und Summenrechnung*, trans. by C. H. Schnuse, Leibrod, Braunschweig 1867.]

— *1865a*: *Treatise on Differential Equations – Supplementary Volume*, ed. Isaac Todhunter, Macmillan, Cambridge.

— *1868a*: "Of Propositions Numerically Definite", in *Transactions of the Cambridge Philosophical Society 11*, 1871, 396-411. Also in *1952a*, 167-186. (Read posthumously by De Morgan 1868.)

— *1952a*: *Studies in Logic and Probability*, ed. Rush Rhees, Open Court, La Salle, Ill. and Watts, London.

Boole, M. *1931a*: *Collected Works*, 4 vols., ed. E. M. Cobham and E. S. Dummer, Daniels, London.

Bornet, G. *1995a*: "George Boole's Linguistic Turn and the Origins of Analytical Philosophy", *The British Tradition in 20th Century Philosophy*, ed. J. Hintikka and K. Puhl, Hölder-Pichler-Tempsky, Wien, 236-248.

— *1995b*: "Booles (Zeichen-)Modell des Geistes – Die metaphorische Wurzel der algebraischen Logik", *Metapher und Innovation*, ed. L. Danneberg, A. Graeser and K. Petrus, Haupt, Bern, 246-267.

Brock, W. H. *1967a*: (Ed.) *The Atomic Debates*, Leicester University Press, Leicester.

Bryant, S. *1888a*: "On the Nature and Functions of a Complete Symbolic Language", *Mind 13*, 188-207.

Buickerood, J. G. *1985a*: "The Natural History of the Mind: Locke and the Rise of Facultative Logic in the Eighteenth Century", *History and Philosophy of Logic 6*, 157-190.

Cayley, A. *1848a*: "Démonstration d'une théorème de M. Boole concernant les intégrales multiples", in *Journal des mathématiques pures et appliquées* (1) *13*, 243-248. Also in *The Collected Mathematical Papers*, Cambridge at the University Press, vol. 1, 1889, 384-387.

— *1853a*: "Note on a Question in the Theory of Probabilities", in *Philosophical Magazine* (4) *6*, 1853, 259. Also in *The Collected Mathematical Papers*, Cambridge at the University Press, vol. 2, 1889, 103-104†.

— *1862a*: "On a Question in the Theory of Probabilities", in *Philosophical Magazine* (4) *23*, 1862, 361-365. Also in *The Collected Mathematical Papers*, Cambridge at the University Press, vol. 5, 1892, 80-84†.

Corcoran, J. *1986a*: "Correspondence without Communication", *History and Philosophy of Logic 7*, 65-75. [Review of Smith *1982a*.]

Corcoran, J. and Wood, S. *1980a*: "Boole's Criteria of Validity and Invalidity", *Notre Dame Journal of Formal Logic*, 21, 609-638.

Couturat, L. *1901a*: *La Logique de Leibniz d'après des documents inédits*, Alcan, Paris. [Repr. Olms, Hildesheim, 1985.]

— *1903a*: *Opuscules et fragments inédits de Leibniz*, Alcan, Paris.

Deakin, M. *1996a*: "Boole's Mathematical Blindness", *The Mathematical Gazette*, to appear.

De Morgan, A. *1836a*: "Calculus of Functions", *Encyclopaedia Metropolitana*, vol. 2, 305-392. [Date of offprint; volume carries "1845".]

— *1847a*: *Formal Logic*, Walton and Maberly, London. [2nd. ed., ed. A. E. Taylor, Open Court, La Salle, 1926.]

— *1850a*: "On the Symbols of Logic, the Theory of the Syllogism, and in Particular of the Copula, and the Application of the Theory of Probabilities to some questions of Evidence", in *Transactions of the Cambridge Philosophical Society 9*, Part 1, 1851, 79-127. (Read 1850). Partly (without pages 116-125) as "On the Syllogism: II." also in *1966a*, 22-68†.

— *1858a*: "On the Syllogism, No. III, and on Logic in General", in *Transactions of the Cambridge Philosophical Society 10*, 1864, 173-230. (Read 1858.) Also in *1966a*, 74-146.

— *1860a*: "Logic", in *English Cyclopedia* (Arts and Sciences), vol. 5, 150-154. Partly also in *1966a*, 247-270†.

— *1867a*: "A Note on Professor Boole's Papers", dated November 30th 1867, typescript R.1 (part of MS 782) in the Library of the Royal Society.

— *1966a*: *On the Syllogism and other Logical Writings*, ed. P. Heath, Routledge & Kegan Paul, London.

Dessì, P. *1988a*: "Editor's Introduction", in Whately *1826a*, reprint, x-xxix.

Diagne, S. B. *1989a*: *Boole – L'oiseau de nuit en plein jour*, Berlin, Paris. [Illustrated biography.]

Dirichlet, J. P. G. *1839a*: "Sur une nouvelle méthode pour la détermination des intégrales multiples", *Comptes rendus de l'Académie des Sciences 8*, 156-160. Also in *Gesammelte Werke*, ed. L. Fuchs and L. Kronecker, vol. 1, Reimer, Berlin 1889; repr. Chelsea, New York, 1969, 375-380†.

Dudman, V. H. *1976a*: "From Boole to Frege", in *Studien zu Frege – Studies on Frege*, vol. 1, (ed.) Schirn, M., Fromman-Holzboog, 109-138.

Durand, M.-J. *1990a*: "Genèse de l'algèbre symbolique en Angleterre: une influence possible de J. Locke", *Revue d'histoire des sciences 48*, 129-180.

Edwards, A. W. F. *1989a*: "Venn Diagrams for Many Sets", *The New Scientist 7*, January, 51-56.

Ellis, R. L. *1863a*: "Notes on Boole's Laws of Thought", in his *The Mathematical and other Writings*, ed. W. Walton, Deighton, Bell, Cambridge, 391-394. Also in *Reports of the British Association for the Advancement of Science*, 1870 (publ. 1871), 12-14.

* Euclid: *Elements*. [Boole probably used a standard edition of his time. On page 177 he cites the edition by François Peyrard: *Les œuvres d'Euclide*, 3 vols., 1814-18, Patris, Paris.]

Feys, R. *1956a*: "Boole as a Logician" and "Boole's Methods of Development and Interpretation", *Transactions of the Royal Irish Academy 57*, 97-112.

Frege, G. *1879a*: *Begriffsschrift, eine der arithmetischen nachgebildete Formelsprache des reinen Denkens*, Nebert, Halle a. S. [Translated in van Heijenoort, J. (ed.), *From Frege to Gödel – A Source Book in Mathematical Logic 1879-1931*, Cambridge Mass., 1-82.

— *1880-81a*: "Booles rechnende Logik und die Begriffsschrift", *Nachgelassene Schriften – Erster Band*, H. Hermes, F. Kambartel, F. Kaulbach (eds.), Meiner, Hamburg 1969, 9-59. [Translation by P. Long and R. White in *Posthumous Writings*, Basil Blackwell, Oxford 1979.†]

— *1884a*: *Grundlagen der Arithmetik*, Koebner, Breslau. [Translated by Austin, J. L., *The Foundations of Arithmetic: A Logico-mathematical Enquiry into the Concept of Number*, Blackwell, Oxford 1950; also Harper, New York 1960.]

— *1918-19a*, "Der Gedanke – Eine logische Untersuchung", *Beiträge zur Philosophie des deutschen Idealismus 1*, 58-77. [Translated by A. M. and M. Quinton "The Thought: A Logical Inquiry", *Mind 61*, 1956, 289-311, also in P. F. Strawson (ed.), *Philosophical Logic*, Oxford University Press, London 1967, 17-38.]

Gratry, A. J. A. *1944a*: *Logic*, trans. and int. by H. and M. Singer, Open Court, La Salle. [Of 5th ed. (1868) of *Philosophie — Logique*.]

Grattan-Guinness, I. *1977a*: "The Gergonne Relations and the Intuitive Use of Euler and Venn Diagrams", *International Journal of Mathematical Education in Science and Technology 8*, 23-30.

— *1982a*: "Psychology in the Foundations of Logic and Mathematics: the Cases of Boole, Cantor and Brouwer", *History and Philosophy of Logic 3*, 33-53. Also in *Psicoanalisi e storia della scienza*, Olschki, Florence, 1983, 93-121.

— *1988a*: "Living Together and Living Apart: on the Interactions Between Mathematics and Logics from the French Revolution to the First World War", *South African Journal of Philosophy 7*, no. 2, 73-82.

— *1990a*: *Convolutions in French Mathematics, 1800-1840. From the Calculus and Mechanics to Mathematical Analysis and Mathematical Physics*, 3 vols., Birkhäuser, Basel; and Deutscher Verlag der Wissenschaften, Berlin.

— *1991a*: "The Correspondence between George Boole and Stanley Jevons, 1863-1864", in *History and Philosophy of Logic 12*, 15-35.

— *1992a*: "Charles Babbage as an Algorithmic Thinker", *Annals of the History of Computing 14*, no. 3, 34-48.

Gregory, D. F. *1839a*: "On the Solution of Linear Differential Equations with Constant Coefficients", *The Cambridge and Dublin Mathematical Journal 1*, 22-32, 378. Also in *1865a*, 14-27.

— *1839b*: "On the Elementary Principles of the Application of Algebraic Symbols to Geometry", *The Cambridge and Dublin Mathematical Journal 2*, 1-9. Also in *1865a*, 150-162.

— *1865a*: *The Mathematical and other Writings*, ed. W. Walton, Deighton, Cambridge.

Hailperin, T. *1981a*: "Boole's Algebra Isn't Boolean Algebra", *Mathematics Magazine 54*, 172-184.

— *1984a*, "Boole's Abandoned Propositional Calculus", *History and Philosophy of Logic 5*, 39-48.

— *1986a*: *Boole's Logic and Probability*, 2nd ed. North Holland, Amsterdam.

Hallam, H. *1829a*: *Introduction to the Literature of Europe, in the Fifteenth, Sixteenth, and Seventeenth Centuries*, 2nd ed., vol. 3, John Murray, London.

Halsted, G. B. *1878a*: "Boole's Logical Method", *Journal of Speculative Philosophy 12*, 81-91.

— *1878b*: "Professor Jevons's Criticism of Boole's Logical System", *Mind 3*, 134-137.

Hamilton, W. *1830a*: "Philosophy of Perception", *Edinburgh Review 52*, 158-207. Also in *1852a*, 39-98. [Discussion of *Œuvres Complètes de Thomas Reid*.]

*— *1833a*: "Logic, in Reference to the Recent English Treatises on that Science", in *Edinburgh Review 4*, 260-277. Also in *1852a*, 117-173†.

*— *1852a*: *Discussions on Philosophy and Literature, Education and University Reform*, Longmans etc., London [2nd ed. 1853, 3rd ed. 1866†].

Hamilton, W. R. *1837a*: "Theory of Conjugate Functions", *Transactions of the Royal Irish Academy 17*, 291-423. Also in *Mathematical Papers*, vol. 1, Cambridge University Press, Cambridge, 1931, 3-96.

Harley, R. *1866a*: "George Boole, F.R.S.", in *British Quarterly Review 44*, 141-181. Also in Boole *1952a*, 425-472†. [Obituary, signed R. H.]

— *1867a*: "George Boole", in *Proceedings of the Royal Society of London 15*, vi-xi. [Obituary, anonymous]

— *1867b*: "Remarks on Boole's Mathematical Analysis of Logic", *Reports of the British Association for the Advancement of Science*, (1866), pt. 2, 3-6.

— *1871a*: "On Boole's Laws of Thought", *Reports of the British Association for the Advancement of Science*, (1870), pt. 2, 14-15.

Heath, P. *1966a*: "Introduction", in De Morgan *1966a*, vii-xxxi.

Hesse, M. *1952a*: "Boole's Philosophy of Logic", in *Annals of Science 8*, 61-81.

Hooley, J. *1966a*: "Boole's Method for Solving Logical Equations", *The Mathematical Gazette 50*, 114-118.

Hughlings, I. P. *1869a*: *The Logic of Names – An Introduction to Boole's Laws of Thought*, James Walton, London. [Despite its title Boole is not the main theme of this book.]

Husserl, E. *1891a*: "Besprechung von E. Schröder, Vorlesungen über die Algebra der Logik", *Göttingsche gelehrte Anzeigen*, 243-278. Also in *Aufsätze und Rezensionen (1890-1910)* (= Husserliana XXII), Nijhoff, The Hague, Boston, London 1979, 3-43†.

Johnson, W. E. *1892a*: "The Logical Calculus", *Mind n.s. 1*, 3-30, 235-250, 340-357.

Jourdain, P. E. B. *1910a*: "The Development of the Theories of Mathematical Logic and the Principles of Mathematics", part 1, *Quarterly Journal of Pure and Applied Mathematics 41*, 324-352. Repr. in *Selected Essays on the History of Set Theory and Logics (1906-1918)*, ed. I. Grattan-Guinness, CLUEB, Bologna, 1991, 104-132. [Articles on Leibniz and Boole.]

Kant, I. *1787a*: *Critik der reinen Vernunft*, 2nd ed., Hartknoch, Riga.

Kempe, A. B. *1886a*: "A Memoir on the Theory of Mathematical Form", *Philosophical Transactions of the Royal Society of London 177*, 1-70.

Kneale, W. *1948a*: "Boole and the Revival of Logic", *Mind n.s. 57*, 149-175.

Laita, L. *1977a*: "The Influence of Boole's Search for a Universal Method in Analysis on the Creation of his Logic", *Annals of Science 34*, 163-176.

— *1979a*: "Influences on Boole's Logic: the Controversy between William Hamilton and Augustus De Morgan", *Annals of Science 36*, 45-65.

— *1980a*: "Boolean Algebra and its Extra-logical Sources: the Testimony of Mary Everest Boole", *History and Philosophy of Logic 1*, 37-60.

Latham, R. G. *1847a*: *First Outlines of Logic Applied to Grammar and Etymology*, London. [Pamphlet]

* Leibniz, G. W. *1840a*: *God. Guil. Leibnitii opera philosophica quae exstant Latina Gallica Germanica omnia*, Part. 1. (Ed.) J. E. Erdmann, Eichler, Berlin 1840.

* — *1681a*: "Addenda ad specimen calculi universalis", in Leibniz *1840a*, 98-99†.

Liard, L. *1877a*: "La logique algébrique de Boole", *Revue philosophique 4*, 285-317.

— *1878a*: *Les logiciens anglais contemporains*, Baillière, Paris.

Lotze, R. H. *1880a*: "Anmerkung über logischen Calcül", *Logik*, 2. ed., Leipzig, 256-269 [Engl. transl. Oxford, 2nd ed. 1888, page 277].

* Lubbock, J. W. *1838a*: *Remarks on the Classification of the Different Branches of Human Knowledge*, London.

* — *1838b*: "On the Divergence of the Numerical Coefficients of Certain Inequalities of Longitude in the Lunar Theory", in *Philosophical Magazine* (3) *12*, 168-172.

* — *1840a*: "On the Heat of Vapours and on Astronomical Refraction", in *Philosophical Magazine* (3) *16*, 434-441, 510-514, 562-569; *17* 272-280, 467-473, 488-507.

Lubbock, J. W. and Bethune, J. E. Drinkwater *1830a*: *A Treatise on Probabilities*, London. Reprinted 1844. (Library of Useful Knowledge.)

MacColl, H. *1877a*: "The Calculus of Equivalent Statements and Integration Limits", *Proceedings of the London Mathematical Society* (1) *9*, 9-20.

Macfarlane, A. *1879a*: *Principles of the Algebra of Logic, with Examples*, Douglas, Edinburgh.

MacHale, D. *1985a*: *George Boole. His Life and Work*, Boole Press, Dublin.

⋇ Mansel, H. L. *1860a*: *Prolegomena Logica· An Inquiry into the Psychological Character of Logical Processes* 2nd ed., H. Hammans, Oxford 1860.

Merrill, D. D. *1990a*: *Augustus De Morgan and the Logic of Relations*, Kluwer, Dordrecht.

Moore, G. H. *1993a*: "Introduction", in *The Collected Papers of Bertrand Russell Vol. 3*, Routledge, London, xiii-xlviii.

Neil, S. *1865a*: "Modern Logicians – The Late George Boole", *The British Controversialist and Literary Magazine, n.s. 80*, 81-94, 161-174.

Newton, I. *1707a*: *Arithmetica universalis sive de compositione et resolutione arithmetica liber*, ed. W. Whiston, 1st. ed. Cambridge. English transl. J. Raphson, 1st. ed. London 1720. [Written in lecture form in the period 1673-1683.]

Øhrstrøhm, P. *1985a*: "W. R. Hamilton's View of Algebra as the Science of Pure Time and His Revision of This View", *Historia Mathematica 12*, 45-55.

Olson, R. *1975a*: *Scottish Philosophy and British Physics 1750-1880*, Princeton University Press, Princeton.

Panteki, M. *1992a*: *Relationships between Algebra, Differential Equations and Logic in England 1800-1860*, Council for National Academic Awards (London), Dissertation.

— *1993a*: "Thomas Solly (1816-1875): an Unknown Pioneer of the Mathematicization of Logic in England, 1839", *History and Philosophy of Logic 14*, 133-169.

Parshall, K. H. *1989a*: "Towards a History of Nineteenth-Century Invariant Theory", in *History of Modern Mathematics*, vol. 1, ed. D. Rowe and J. McCleary, Academic Press, New York, 157-208.

Peacock, G. *1834a:* "Report on the Recent Progress and Actual State of Certain Branches of Analysis", *Reports of the British Association for the Advancement of Science*, (1833), 185-332.

Peckhaus, V. *1995a*: *Hermann Ulrici (1806-1884). Die Hallesche Philosophie und die englische Algebra der Logik*, Hallescher Verlag, Halle.

— *1996a*: "Leibniz und die britischen Logiker des 19. Jahrhunderts", in *Leibniz und Europa*, ed. H. Breger, Steiner, Stuttgart, to appear.

Peirce, Ch. S. *1880a*: "On the Algebra of Logic", *American Journal of Mathematics 3*, 15-57. Also in *Writings of Charles S. Peirce – Volume 4 (1879-1884)*, Ch. Kloesel (ed.), Indiana University Press, Bloomington and Indianapolis 1986, 163-209.

Prior, A. N. *1949a*: "Categoricals and Hypotheticals in George Boole and his Successors", *Australian Journal of Philosophy 27*, 171-196.

Rhees, R. see also Boole *1952a*.

— *1952a*: "Note in Editing", in Boole *1952a*, 9-43.

— *1955a*, "George Boole as Student and Teacher – By some of his Friends and Pupils", *Proceedings of the Royal Irish Academy 57* (1954-1956), sec. A., 74-78.

Rice, A. *1996a*: "Augustus De Morgan: Historian of Science", *History of Science 34*, 201-240.

Richards, J. *1980a*: "Boole and Mill: Differing Perspectives on Logical Psychologism", *History and Philosophy of Logic 1*, 19-36.

— *1980b*: "The Art and the Science of British Algebra: A Study in the Perception of Mathematical Truth", *Historia Mathematica 7*, 343-365.

Ross, G. R. T. *1905a*: "On Verb-functions, With Notes on the Solution of Equations by Operative Division", *Proceedings of the Royal Irish Academy 25*, sec. A, 31-76.

Royal Society *1893a*: *Minutes of the Council of the Royal Society from October 30th, 1884 to June 30th, 1892*, Harrison, London.

— *1899a*: *Minutes of the Council of the Royal Society from October 30th, 1892 to June 30th, 1898*, Harrison, London.

Russell, B. *1901a*, "Recent Work on the Principles of Mathematics", *International Monthly 4*, 1901, 83-101. Also in *The Collected Papers of Bertrand Russell Vol. 3*, G. H. Moore (ed.), Routledge, London 1993, 366-379†.

— *1903a*: *The Principles of Mathematics*, Cambridge University Press [2nd. ed, George Allen & Unwin, London, 1938]

Ryall, J. *1865a*: "Obituary notice (George Boole)", in *Illustrated London News 21*, January, 59, 61.

Scanlan, M. J. *1991a*: "Who were the American Postulate Theorists?", *The Journal of Symbolic Logic 56*, 981-1002.

Skidelsky, R. *1983a*: *John Maynard Keynes*, vol. 1, MacMillan, London.

Smee, A. *1851a*: *The Process of Thought Adapted to Words and Language – Together with a Description of the Relational and Differential Machines*, Longman, Brown, Green, and Longmans, London.

Smith, G. C. *1982a*: (Ed.) *The Boole – De Morgan Correspondence 1842-1864*, Clarendon Press, Oxford.

— *1983a*: "Boole's Annotations on 'The Mathematical Analysis of Logic'", *History and Philosophy of Logic 4*, 27-38.

Styazhkin, N. I. *1969a*: *History of Mathematical Logic from Leibniz to Peano*, MIT Press, Cambridge, Mass.

* Suarez, F. *1605a*: *Metaphysica disputationum ...*, 2 vols. Baltharus Lippsius, Moguntiae [= Mainz].

Terrot, Ch. H. *1857a*: "On the Possibility of Combining Two or More Probabilities of the Same Event, so as to Form One Definite Probability", in *Transactions of the Royal Society of Edinburgh 21*, 369-376. Also in Boole *1952a*, 487-496.

van Evra, J. *1977a*: "A Reassessment of George Boole's Theory of Logic", *Notre Dame Journal of Formal Logic 18*, 363-377.

— *1984a*: "Richard Whately and the Rise of Modern Logic", *History and Philosophy of Logic 5*, 1-18.

Ulrici, H. *1855a*: [Review of Boole *1854a*], *Zeitschrift für Philosophie und philosophische Kritik N.F. 27*, 273-291. Also in Peckhaus *1995a*, 87-104.

Vassallo, Nicla *1995a*: *La Depsicologizzazione della Logica – Un confronto tra Boole e Frege*, FrancoAngeli.

Venn, J. *1881a*: *Symbolic Logic*, London [2nd ed. revised and rewritten 1894]

Vercelloni, L. *1989a*: *Filosofia delle strutture*, La Nuova Italia Editrice, Florence.

Whately, R. *1823a*: "Logic", in *Encyclopædia metropolitana*, vol. 1, 193-240.

* — *1826a*: *Elements of Logic Comprising the Substance of the Article in the Encyclopædia Metropolitana with Additions, &c.*, Mawman, London 1826. [Reprinted with introduction by P. Dessì, Cooperativa Libraria Universitaria Editrice Bologna, Bologna, 1988.]

* — *1856a*: *Reviews* "On Instinct", in *Miscellaneous Lectures and reviews*, Parker, Son and Brown, London, 60-84.

Wood, S. *1976a*: *George Boole's Theory of Propositional Forms*, State University of New York at Buffalo, Dissertation.

Young, G. P. *1865a*: "Remarks on Professor Boole's Mathematical Theory of the Laws of Thought", *The Canadian Journal 10*, 161-182.

Index of Names

Index of Subjects

Mathematics with Birkhäuser

PHILOSOPHY • LOGIC • MATHEMATICS

F.A. Rodríguez-Consuegra, University of Valencia, Spain

Kurt Gödel

Unpublished Philosophical Essays

1995. 236 pages. Hardcover
ISBN 3-7643-5310-4

Kurt Gödel, together with Bertrand Russell, is the most important name
in logic, and in the foundations and philosophy of mathematics of this
century.

He devoted more years of his life to philosophy than to technical investigation, wrote hundreds of
pages on the philosophy of mathematics, as well as on other fields of philosophy. Reluctant to de-
vulge his ideas even to close friends, it was only possible to learn more about them after the opening of his literary estate at Princeton a decade ago. The goal of this book is to make available to the
scholarly public solid reconstructions and editions of two of the most important essays which Gödel
wrote on the philosophy of mathematics.

The book is divided into two parts. The first provides the reader with an incisive historico-philoso-
phical introduction to Gödel's technical results and philosophical ideas. The second contains two of
Gödel's most important and fascinating unpublished essays: 1) the Gibbs Lecture ("Some basic theorems on the foundations of mathematics and their philosophical implications", 1951), where Gödel
describes the essentials of his famous metamathematical results and discusses some of the philosophical implications; and 2) two of the six versions of the essay which Gödel wrote for the Carnap
volume of the Schilpp series The Library of Living Philosophers ("Is mathematics syntax of language?",
1953–1959), but which was never actually submitted.

Dr. Rodríguez-Consuegra is a bright new light in the study of mathematical logic, set theory, and
the philosophy of mathematics as these developed over the past twelve decades.... With the
present book he establishes yet another milestone in Spain's impressive latter-day progress in
scientific philosophy.
From the foreword by W.V. Quine

*"...The volume is handsomely produced, and will be essential reading for any [...] metaphysician or
philospher of mathematics."*

AUSTRALIAN JOURNAL OF PHILOSOPHY, 9/1996

For orders originating from all over
the world except USA and Canada:
Birkhäuser Verlag AG
P.O Box 133
CH-4010 Basel/Switzerland
Fax: +41/61/205 07 92
e-mail: farnik@birkhauser.ch

For orders originating in the
USA and Canada:
Birkhäuser
333 Meadowland Parkway
USA-Secaurus, NJ 07094-2491
Fax: +1 201 348 4033
e-mail: orders@birkhauser.com

 Birkhäuser

Birkhäuser Verlag AG
Basel · Boston · Berlin

VISIT OUR HOMEPAGE **http://www.birkhauser.ch**